2nd Edition

Computational Algebra

Sun Tae Soh

Department of Mathematics
Myong Ji University

KYOWOO

Kyowoo Co., Ltd.
8, Yangnyeongsi-ro, Dongdaemun-gu,
Seoul, Korea 130-861,

Phone: 82-2-925-2861
F a x: 82-2-925-2860
e-mail: kyowoo@kyowoo.co.kr
http://www.kyowoo.co.kr

Printed in the Republic of Korea.
ISBN 979-11-251-0316-5 (93410)

₩ 24,000

To my parents for their love and sacrifice in raising us, four sons.

Contents

Preface to the Second Edition

This book, *Computational Algebra*, was first published in 2006 and then was improved in the following years for some time. Since then more time has passed fixing some typos and errors of mine as much as I could, I restated more clearly or added, if necessary, several statements of the previous book, hoping to improve its readability. Thus, this edition would be called the final version of my book, *Computational Algebra.*

As a remark of this preface, I would like to mention briefly what I had foreseen in 1999, 21 years ago, what will be the main aspect of Mathematics in the 21st century, that is included as an Appendix A: An Introduction to Algebraic Analysis,[1] of my book, because most of the people around the world nowadays recognize the importance of Computation Related Human Activities(CRHA), after watching AlphaGo's victory in a Go-game competition in Seoul, in the Spring of 2016. Since the main aspect of CRHA consists of (1) Symbolic Computation, i.e., Algebraic Analysis, (2) deductive reasoning or inductive reasoning for intellect's understanding, and (3) decision making process in order to reach a meaningful conclusion after finitely many steps in reasoning process.

I think it is very important for us, human-beings, to understand what each of these three actually means, and it is fairly well described in the Appendix A: An Introduction to Algebraic Analysis, of my book, although I did not mention them explicitly in terms of technical terminologies, because, as it is well-known that the Greek Philosophy of those philosophers such as Socrates (470 - 399 BC), Plato (428/427 or 424/423 – 348/347 BC), and Aristotle (384 - 322 BC) is based on the method of Mathematical proof, still nowadays only well-taught in the courses of Pure

[1] http://ch.catholic.or.kr/pundang/4/soh/mathnet_973071.htm

Mathematics, which had already well been developed by the Pythagorean School of 6th century BC, CRHA are also definitely based on the method of Mathematical proof, still nowadays only well-taught in the courses of Pure Mathematics.

I believe that people well know that there is no contradiction among any facts, as long as they are all Mathematically proved to be true, and hence this is the main reason why the method of the Mathematical proof is very important to us, human-beings, as long as we reason why. Another thing is that people usually do not make a mistake in deriving a true conclusion through a deductive reasoning from sound hypotheses, but people sometimes derive a false conclusion through an inductive reasoning even from sound hypotheses that is NOT a mathematical inductive reasoning. One should also recognize that Artificial Intelligence is born to have such a capacity but it inevitably has, like us human-beings, a defect in its inductive reasoning process.

I think the Appendix A: An Introduction to Algebraic Analysis and the first chapter of my book is in fact well-suited for the beginners who want to learn how to better think and understand CRHA in the Computer Age, the 21st century, because their main underlying theme is not about any particular subjects in Mathematics, but rather about how we should think and reason to derive a meaningful conclusion in an axiomatic manner.

Toward this direction, some amount of my research works and papers on finding or establishing a new relationship among objects, i.e., concepts/ideas, in Computation Related Human Activities(CRHA), that are meaningful to us human-beings, existing generically or specially among the given ideas/concepts, under the environment of internet-based Artificial Intelligence, which have been produced by the author for a period of more than recent 15 years, are available to the readers through the internet access.[2]

Sun T. Soh

March 2020

[2]http://ch.catholic.or.kr/pundang/4/soh/a_list_of_research_outputs.htm

The rest of the preface is the original preface to the first edition of this book:

Preface [to the First Edition]

We recall that we extend our number system, the set $\mathbb{N} = \{1, 2, 3, ...\}$ of all natural numbers, to the set $\mathbb{Z}$ of all integers from our sincere efforts of obtaining, for instance, a solution for X for a given linear equation, $X + n = 0$ for a given number $n \in \mathbb{N}$. Here, we note that the word, natural, means "God-given".

This is perhaps our original source for the right definition of -3 as "a (new, I mean, non-natural) number whose addition to the given number 3 results in the 0, i.e., emptiness", i.e. $(-3) + 3 = 0$. Of course, we "know" that we have only one such a new number if ever it exists within our domain of thoughts. Thus, given a number 3, "*a linear relation of being equal to each other*", $(-3) + 3 = 0$, exactly tells us that we "know" either -3 from 0 or 0 from -3, as long as we live in $\mathbb{Z}$, i.e., we accept these (non-natural numbers) as new members of our numbers.

Similarly, we extend our number system, $\mathbb{Z}$, to the set $\mathbb{Q}$ of all rational numbers from our sincere efforts of obtaining, for instance, a solution for X for a given linear equation, $nX = 1$ for a given non-zero number $n \in \mathbb{Z}$. Similarly, this process thus allows us to have the right definition of the new number $\frac{1}{3}$. It is a number whose multiplication to the given number 3 results in 1, i.e., the unit. Of course, we have only one such a new number if ever it exists within our domain of thoughts. Thus, given a number 3, "*a linear relation of being equal to each other*", $\frac{1}{3} \cdot 3 = 1$, exactly tells us that we "know" either $\frac{1}{3}$ from 1 or 1 from $\frac{1}{3}$, as long as we live in $\mathbb{Q}$, i.e., we accept these (another non-natural numbers) as new members of our numbers. The reason why we mention these very basic stuffs is that it in fact allow us to understand what is Algebraic Analysis, or equivalently, Symbolic Computation.

For instance, 0.3333333333 is never the same as $\frac{1}{3}$ since

$$0.3333333333 \cdot 3 = 0.9999999999 \neq 1.$$

In Numerical Analysis, however, we simply approximate $\frac{1}{3}$ as 0.3333333333 and we regard 0.3333333333 as $\frac{1}{3}$ in all later computation, although it is mathematically absolutely wrong to treat 0.3333333333 in actual computation as if it were $\frac{1}{3}$.

Why did we accept this? Perhaps this has been the case because Numerical Analysis was studied extensively before enough development of software technologies especially for Symbolic Computation. Consequently, most students and professors in Engineering College still do not know or even understand the main differences between Numerical Analysis and Algebraic Analysis, i.e., Symbolic Computation. One of these two subjects does not replace the other. Rather, each compensates the other's weak points, so that they help each other.

By studying this book, we strongly hope the readers to learn the main points of Algebraic Analysis which should eventually lead the readers to recognize how much these two subjects, Numerical Analysis and Algebraic Analysis, are different.

Even when

(1) multiple precision is guaranteed and
(2) rational numbers such as $\frac{1}{3}$ are recognized by computing machines like computers mathematically properly which are definitely necessary for Algebraic Analysis,

the subject Algebraic Analysis is divided into two: Linear Algebra or Nonlinear Algebra, depending on whether multiplication among unknown variables is not allowed or allowed.

What are we supposed to do when we are given a system of linear equations in several unknowns? We just need to learn how to solve such a linear system and this is all about Linear Algebra. Thus Linear Algebra will tell us what's really going on when we attempt to solve a system of linear equations. For instance, once you learn this subject a little bit, you can tell others explicitly and affirmatively whether such a system has a solution or not, and if so, you can actually tell others about how to compute all the solutions of the system in a systematic manner. Understanding this part of Linear Algebra is very important because almost every subject in Linear Algebra can be expressed as a system of linear equations.

Then what about a system of non-linear, i.e., polynomial equations in several unknowns? Our understanding is that virtually no books on either Linear Algebra or Algebra attempt to mention this, and consequently most students who take Linear Algebra or Algebra courses never have a fair chance of thinking about its usefulness and importance in advance.

In the main treatment of this book, we try to cover all the basic materials for non-Linear Algebra, i.e., multiplication-related computational aspects with "one" unknown as much as we could. Unlike other books in Algebra, we start our discussion axiomatically, but as we go further, we try hard to move toward computational aspects of the subjects. Here are the contents of each chapter of this book.

This book has grown out of my teachings on the subject for several years in my University. In writing this book, I took M. Artin's famous book, *Algebra*, as a basic model. On the other hand, this book is also prepared as an introduction to the book, *Algebraic Structures*, of my teacher, late Professor George R. Kempf (1944 - 2002) at JHU, with a strong computation-oriented taste. For example, in Chapter 6 of this book, the reader will find some useful examples as an application of the Theory of Invariants of the 19th century and the Theory of Groebner basis of the 20th century.

In Chapter One, we introduce the most basic and hence common terminologies in Algebra with some motivational explanations. Since the materials are treated in an axiomatic manner, it is helpful for the beginners to have a chance of looking at the bare skeleton of the whole subject as roughly as possible, which is the main purpose of preparing our chapter one.

Thus, Chapter 2 is the real beginning of this book. In this chapter, we try to study the reverse process of multiplication process in general, i.e., factorizability of numbers and polynomials in one variable, and thus it is basically all about the polynomial rings in one variable. In particular, the distinction between prime elements and irreducible elements in an integral domain is clearly made through several examples even at the beginning of the Chapter 2. The materials in this chapter is essential to later chapters of this book.

In Chapter 3, we study another kind of factorization in an advanced Linear Algebra, based on generators and relations among them of finitely generated R-modules. The main computing means of this chapter is the diagonalization process of integer matrices. In particular, we prove the structure theorem for abelian groups. We also discuss rational canonical forms and Jordan canonical forms in this chapter.

In Chapter 4, we study field extensions in general, based on the materials we covered in Chapter 2 of this book. In particular, we briefly mention how we can strengthen the method of undetermined coefficients as an application of the Theory of Groebner basis. See, for instance, Example 4.10, Example 4.12, Example 4.13, Example 4.14, Example 5.5, and Example 5.8.

In Chapter 5, we discuss Galois theory in characteristic 0 case. Since we are dealing with solution sets of polynomials over a field, this chapter basically requires us to compute in an exact manner. In particular, we also apply the Theory of Groebner basis to actually compute the minimal polynomial of a primitive element in a finite extension of $\mathbb{Q}$. For this, see Example 5.5.

In Chapter 6, we discuss the important role of the discriminant of an irreducible polynomial $f(x)$ over $\mathbb{Q}$ in determining the Galois group of $f(x)$ over $\mathbb{Q}$, and we apply the Theory of Invariants and the Theory of Groebner basis in computing the discriminant explicitly. We shall see how the combination of these two theories can actually help us in dealing with computation-related arguments, and several computed examples are provided in this chapter. See, for instance, Example 6.6, Example 6.9, Example 6.10, Example 6.14, Example 6.16, and Example 6.17.

We thus believe that Chapters 4, 5, and 6 of this book are particularly well prepared for a smooth introduction to the subject, called **Algebraic Analysis**, that is, symbolic computation.

Unless we approximate numbers, which is very common in Numerical Analysis, the nature of all the objects under our mathematical consideration is exact. However, this in general requires us to multiply more than one unknowns and thus this requires us to study a polynomial ring in several (i.e., more than one) variables over an (algebraically closed) field, called "Computational Algebraic Geometry", which is beyond the scope of the main theme of this book. On the other hand, since even Personal Computers nowadays already provide us with an excellent Hardware means for symbolic computation and since this theory provides us with a surprising means for solving a system of polynomial equations, we add many computed examples which require a Computer Algebra System, REDUCE, and these examples should convince the readers of the importance and usefulness of symbolic computation, or Algebraic Analysis.

For more about Computer Algebra Systems, one may visit our homepage:

http://trinitas.mju.ac.kr/intro2softwares.html .

There, you will further find very useful information for users on CAS, Maple, Reduce, Matlab, etc.[3] Also,

http://trinitas.mju.ac.kr/intro2InetCompu.html

describes an important aspect of the availability of Internet environment to everybody in this wonderful world of 21st century.[4]

[3] http://ch.catholic.or.kr/pundang/archive/4/intro2softwares.html
[4] http://ch.catholic.or.kr/pundang/archive/4/intro2InetCompu.html

CHAPTER 1

A Few Fundamental Concepts in Algebra

1. Introduction

In this introductory chapter, we are going to review a few very basic concepts and definitions in Algebra. Before we start what constitutes Algebra, we would like to have a look at what had happened in our numbers, first. The main purpose of this chapter is for the readers eventually to have a good sense in Symbolic Computation or Algebraic Analysis. Another reason is that quite many axioms used in defining algebraic objects are in fact those satisfied by natural numbers, integers, or rational numbers.

2. Extensions From Natural Numbers to Complex Numbers

Every human being knows what are natural numbers, otherwise he is not one of us. That is, we know that if we add the number one to the number one, then we have the number two, and if we add the number one to the number two, then we have the number three, etc., i.e., we can keep going on this indefinite process to obtain another natural number repeatedly. This is our first incidence of recognizing an infinite set, the set of all natural numbers,

$$\mathbb{N} = \{1, 2, 3, 4, ...\},$$

and nobody knows exactly why and how we have recognized natural numbers in this way.

Moreover, we "know" that every non-empty subset S of $\mathbb{N}$ has a least element in it, i.e., we "accept" this as a fact without its proof. We also "know" that **the associative laws**

$$a + (b + c) = (a + b) + c \quad \text{and} \quad a \cdot (b \cdot c) = (a \cdot b) \cdot c.$$

hold for both addition and multiplication of any three natural numbers a, b, and c. We also "know" that **the distributive laws**

$$a \cdot (b + c) = a \cdot b + a \cdot c \quad \text{and} \quad (a + b) \cdot c = a \cdot c + b \cdot c$$

hold for natural numbers, etc. This means that when we say that we "know" natural numbers, we in fact means that we also accept these axioms as facts, and we are ready to apply these axioms in our later arguments on numbers. Since everybody knows about what constitutes natural numbers these numbers are some times called God-given numbers. That is, "natural" means "God-given", and conversely.

Then how about the negative numbers? Suppose a wife has prepared a tonight's dinner for 3 family members, including herself. But, her husband brought 2 friends with him to have dinner together at his house tonight. Then, his wife will quickly figure out that there will be no food for her, and this will make her unhappy. The situation is abstracted as follows. Given the number 3, she basically figures out the relation that

$$(-3) + 3 = 0,$$

where recognizing the zero, 0, or the negative number, -3, requires the recognition of the other. The point here is that she recognizes this relation first in her head; and then the usefulness of the negative number -3 and/or the number 0 (here meaning that nothing left to her). Here, by definition,

DEFINITION 1.1. *0 is a number whose addition to any number n is any number itself.*

Since this kind of arguments is always possible for any other number n, our slow abstraction leads us to the introduction of the set of integers,

$$\mathbb{Z} = \{..., -3, -2, -1, 0, 1, 2, 3, ...\}.$$

Thus, *by definition*, the number -3 is a number whose addition to the natural number 3 results in the number 0 (i.e., nothing left). We then easily check that if there were such a number -3 then it must be unique. But since this number is not an element of $\mathbb{N}$, we need to extend the set $\mathbb{N}$ of natural numbers to the set $\mathbb{Z}$ of integers, so that we can do such an arithmetic.

This can be more abstractly expressed as follows: Solve the following equation for the unknown X:

$$X + 3 = 0.$$

Since its solution -3 is not one of the natural numbers, we had to extend our number system from $\mathbb{N}$ to $\mathbb{Z}$ in order to satisfy arithmetic demands of our everyday life like the above example.

With this information, we can prove that

LEMMA 1.1.

$$-(-1) = 1.$$

PROOF. For, $-(-1)$ exactly means that it is a number whose addition to the number -1 results in the number 0 (i.e., nothing left). That is,

$$-(-1) + (-1) = 0.$$

But, we also know

$$1 + (-1) = 0.$$

Since there is only one number whose addition to -1 results in the number 0, we conclude that $-(-1) = 1$. □

Similarly, suppose we are given a 12-inch round pizza and 6 persons are going to equally share it. Then we obviously know how to divide it into 6 pieces of equal size and equal shape, and this establishes an abstract relation,

$$\frac{1}{6} \cdot 6 = 1,$$

which equivalently says that the addition of the same $\frac{1}{6}$ six times is equal to the number 1, or equivalently it says that $\frac{1}{6}$ is a number whose multiplication to the natural number 6 is the number 1. Thus, by definition $\frac{1}{6}$ is a number whose multiplication to 6 results in the number 1 (i.e., a unit.) Here, by definition, 1 is a number whose multiplication to any number n is any number itself. We then easily check that if such a number exists then it must be unique. Since this kind of arguments is always possible for any other number n, our slow abstraction leads us to the introduction of the set $\mathbb{Q}$ of rational numbers, which is an extension of $\mathbb{Z}$. This can be more abstractly expressed as follows: Solve the following equation for the unknown X:

$$X \cdot 6 = 1.$$

Since its solution $\frac{1}{6}$ is not one of the integers, we had to extend our number system from $\mathbb{Z}$ to $\mathbb{Q}$ in order to satisfy arithmetic division demands of our everyday life like the above example.

Moreover, we have

LEMMA 1.2.

$$0 \cdot 1 = 0.$$

PROOF. This is because, for any number n, we have

$$\begin{aligned} 0 \cdot 1 &= (n + (-n)) \cdot 1 \\ &= n \cdot 1 + (-n) \cdot 1 \\ &= n + (-n) \\ &= 0. \end{aligned}$$

□

On the other hand, Greek mathematicians thought that the set of positive rational numbers must be everything, i.e., there are no other numbers they can think of. But, this was a big mistake. Consider a rectangular triangle of equal sides 1. Then even Babylonians in almost $4,000$ years ago knew (without proof) that there *is* a number whose square is equal to $1+1=2$, from the ruler and compass construction of magnitudes, i.e., positive numbers. This number is represented as the square root of 2:

$$\sqrt{2}.$$

LEMMA 1.3. *$\sqrt{2}$ is not a rational number.*

PROOF. Suppose that it is a rational number, say $\frac{a}{b}$ with $(a,b)=1$, i.e., a and b are relatively prime. Then whose square is two exactly means that $a^2 = 2b^2$; hence a^2 is divisible by 2; hence a itself is divisible by 2; hence $4a'^2 = 2b^2$. This just implies that b is again divisible by 2, which is a contradiction to our choice of a and b being relatively prime to each other. □

This can be more abstractly expressed as follows: Solve the following equation for the unknown X:

$$X^2 = 2.$$

Obviously, the disciples of Pythagoras should know about this proof, but history tells that they were forbidden officially to discuss this strange number, $\sqrt{2}$.

Another famous example is π. By its geometric definition, it is a ratio of two magnitudes, the length of circumference of a circle to that of the diameter of the same circle. Thus, if both of these magnitudes are rationals then π must be a rational number. Thus, the question is: when we choose the diameter of our circle to be the unit 1, the length (i.e., the magnitude) of the circumference of the same circle is whether rational or not.

Of course the right answer to this question took a pretty long time and we nowadays know that π is not a rational number. The history of mathematics tells us that pretty some time ago one mathematician spent his whole life to see if the decimal representation of π is periodic or not. (It is a fact that if its decimal representation is periodic, then it is a rational number.) Of course, he could not prove that it is periodic during his life time since by dividing infinitely many times for decimal fractions, he attempted to prove what he must prove its complete opposite, and another mathematician eventually proved that it is not a rational number at all, spending something like almost more than 20 years of his life time!

We note, unlike other examples as we have seen so far, that π is not accepted as a number as a solution of an equation containing an unknown X.

In fact, it took more than $1,800$ years since the Greek period for us to properly define the real numbers using the concept of limits, or limiting process, and it turns out that among real numbers, there are more irrational numbers (i.e., non-rational numbers) than rational numbers.

Having accepting the real numbers $\mathbb{R}$ as numbers, we still have a difficulty if we are asked to solve an equation for X:

$$X^2 + 1 = 0.$$

Obviously, there are no real solutions to this equation, because a square of any number is positive unless it is zero. On the other hand, if there is a number whose square is -1, then mathematicians in Italy of Renaissance period figured out that a very interesting thing happens.

3. A Few Remarks on Connectives

Before we go further, we would like to discuss how the logical connectives are used. In particular, we will discuss the right way of using implication in mathematics. Suppose we are given two input statements p and q. Then the truth value of $p \wedge q$, $p \vee q$ and $p \Rightarrow q$ as outputs are defined as follows:

p	q	$p \wedge q$	$p \vee q$	$p \Rightarrow q$
T	T	T	T	T
T	F	F	T	F
F	T	F	T	T
F	F	F	F	T

Thus, we know immediately the truth value of the outputs $p \wedge q$ and/or $p \vee q$, once we know whether each of the two inputs p and q are true or false.

But the situation is quite different when the usage of the $p \Rightarrow q$ is involved. Although it is defined as is in the above table, its usage in mathematics is the following:

p	$p \Rightarrow q$	q
T	T	T
T	F	F
F	T	T
F	T	F

Here, the only input is p and we want to verify (or prove) whether q is true or false as an output, through a mathematical proof, which corresponds to the step $p \Rightarrow q$. When our input (or hypothesis) p is true, since the truth value of both $p \Rightarrow q$ and q are exactly the same by definition, our final result (or output) q must be true (false respectively), as long as our argument, $p \Rightarrow q$, is true (false respectively).

p	$p \Rightarrow q$	q
T	T	T
T	F	F

On the other hand, when our input (or hypothesis) p is false, our output q can be either true or false regardless of our argument $p \Rightarrow q$ being true. For instance, any mathematical conclusion from a false hypothesis, like 3 divides 5, could be either true or false regardless of our argument $p \Rightarrow q$ being true.

p	$p \Rightarrow q$	q
F	T	T
F	T	F

Now, here is our explanation on the so-called method of contradiction in our proof business and it goes as follows: Suppose we are able to reach a conclusion q through a true argument, i.e., a valid $p \Rightarrow q$ step, from a hypothesis p whose truth value is not yet verified, but on the other hand everybody is able to see that "that" conclusion q is in fact false, then the **only** conclusion we can draw is that the hypothesis p is false. This is in fact the exact meaning of the following table, where the truth value of p is the output.

p	$p \Rightarrow q$	q
F	T	F

EXAMPLE 1.1. *A good example of a proof by the method of contradiction is the proof of Lemma 1.3 above.*

4. What is Algebra?

The subject Algebra has a long history in Mathematics. For instance, even in Greek period, people knew how to add, subtract, multiply, and divide the **magnitudes**, even if they did not know how to write down numbers efficiently and even if they did not have symbols for these four basic operations, $+, -, \times$, and $\div$. They did all their mathematical thinking in a verbal way, based on the so-called method of "Dialogues". Even with the Roman way of representing numbers people had a great difficulty in multiplying numbers. As time goes, people very slowly began to introduce several efficient symbols to represent numbers and the basic four operations on numbers, etc.

When the followings are what we are about to deal with, we call it **Algebra**.

- We treat both **numbers** such as $1, 2, 3, ...$, and **symbols** such as $a, b, c, \pi, x, y, z, ...$
- Some of the symbols, for instance, $a, b, c, ...$ etc., are already **known** to us and others, for instance, $x, y, z, ...$etc., are yet **unknown** to us, but they are treated as if they were numbers, regardless of being known or unknown to us.
- These numbers and symbols are connected using the four connectives: $+, -, \cdot$, and $\div$ (**Addition, Subtraction, Multiplication, and Division**, respectively).

- We "**equate**" two such expressions with = (Equality) sign to have an equation.
- We apply the following rules (or principles) to an equation:
  - (**Addition rule**) Whenever we add a number or a symbol on one side of an equality, we have to do the same thing on the other side of the equality. [Whenever we subtract a number or a symbol from one side of an equality, we have to do the same thing on the other side of the equality.]
  - (**Multiplication rule**) Whenever we multiply [divide] a non-zero number or a symbol on one side of an equality, we have to do the same thing on the other side of the equality.
  - (**Substitution rule**) Whenever we substitute a symbol in one side of an equality with another symbol or number, we have to the same thing on the same symbol on the other side of the equality.
- We try to express any unknown symbol to be equal to a **polynomial expression** of the rest of known and unknown symbols and numbers.

REMARK 1.1. *We remark that the* ***decimal notation*** *of numbers such as* 12345 *and those* ***connecting symbols*** *such as* $+, -, \cdot, \div$, *and* $=$ *were not generally used almost until the end of* 15th *century. Before this time, most mathematical expressions are given as plain sentences, as are in Euclid's Elements. For a long time, the method of the so called "****Dialogues****" has been the best method of learning mathematics.*

5. Why (Abstract) Algebra?

The main objects of our study are polynomials in either one variable or more than one variables over a field k. Only the most elementary mathematical skills are are required to manipulate polynomials, and it is really necessary to work within the larger frame of abstract, i.e. axiomatic, algebra in general.

The concept of abstract algebra arises from the observation that certain operations such as addition and multiplication can be performed on a variety of objects, such as numbers, polynomials, functions, or matrices, to name a few. Certain properties of these operations are often shared by different objects on which these operations are performed.

For example, we consider addition, multiplication, and division with remainder of integers which we will discuss very soon. We will also see that these operations can be performed not only with integers, but with polynomials in one variable over the field of rational numbers or that of real numbers (in fact, over any field), and that the same basis properties such as commutativity or the associativity hold.

Thus, if two sets share a structure and a property of that structure, then they also share all consequences of this property, and hence we would like to derive these consequences in a generalized setting. For example, all the results on integers we will discuss will have precise counter parts for polynomials in one variable, and this is exactly what abstract (i.e., axiomatic) algebra is all about.

6. What is Linear Algebra?

It is certainly a branch of Algebra with only one restriction: We **never** multiply unknown symbols or quantities together (which usually results in more unknown symbols). This means that linear algebra is about how to solve a system of linear equations, and it is about how to use such a computational method in handling other parts of mathematics as applications. Whenever this restriction is applied, it belongs to the subject, called **Linear Algebra**. Since computation itself has no taste at all, we usually need to understand what's really going on behind such computations. This means that we need to learn new concepts which allow us to study certain important properties and/or results in Linear Algebra.

But, as Chapter 3 of this book indicates, certain advanced results in Linear Algebra require some non-linear algebra stuffs.

7. What is Non-Linear Algebra?

On the other hand, the existence of a subject called Linear Algebra then also means the existence of a non-linear Algebra in general, where several unknown symbols or quantities are allowed to be multiplied together. This means that the business of doing Linear Algebra is conceptually included in that of doing Non-Linear Algebra.

For example, if X is unknown, then X^2 is another unknown that is more difficult to know than X itself, and X^3 is another unknown that is more difficult to know than X^2, and so on. The degree of difficulty may be seen graphically by drawing, for instance, the graphs of $y = x$, $y = x^2$, $y = x^3$, etc.

But the situation was changed dramatically at around the middle of the 20th century. Starting 1965 we are now able to solve a system of polynomial equations, thanks to Austrian Mathematician **Bruno Buchberger**'s discovery of an algorithm for computing the so-called "Groebner Basis" for a polynomial ideal in several variables over a field k.

In fact, his discovery of an algorithm for computing a Groebner basis is very important: It is a natural generalization of the algorithmic process of obtaining the row-reduced echelon form of a system of linear equations in Linear Algebra, and it is also somewhat a natural generalization of either finding the greatest common divisor of two polynomials in one variable or that of two natural numbers. Moreover, his theory is in fact a generalization of Gauss' famous result, known as "**the fundamental theorem of Algebra**":

THEOREM 1.1. *(Gauss) Every polynomial equation of degree d in one variable over the field $\mathbb{C}$ of complex numbers has exactly the degree d number of complex roots.*

This may need at least some further explanations.

When we are given a system of finite number of polynomial equations in one unknown, i.e., variable, over a field k, we do not need to attempt to solve common roots of these equations. Instead, we first compute their greatest common divisor $d(X)$ and then solve the single equation $d(X) = 0$ to figure out all the possible common roots of them, where the existence of solutions of $d(X) = 0$ is completely guaranteed by the Gauss' theorem above, and for different reasons we sometimes need to apply the famous numerical method, Newton's method, to figure out the solutions of $d(X) = 0$. In a later chapter, we will have a chance to review these elementary facts.

On the other hand, when we are given a system of polynomial equations in more than one unknown over even an algebraically closed field, finding their common solutions is generally very difficult, because the division process involved is very complicated, and here is the place where Buchberger's theory discovered in the year 1965 properly puts its stake. His theory is a natural generalization of the Gauss' theorem above, because it always **algorithmically** allows us to transform the given system of polynomial equations into another system of polynomial equations, of course, having the same solution set, in such a way that if there is no solution, then we can recognize it computationally, or else the last equation of the latter is a polynomial equation in one unknown, and the rest of the equations of the latter system is of the form which allows us to apply the so-called backward substitution process, so that we can now computationally figure out all the solutions of the system of polynomial equations over an algebraically closed field k. It took about $2,500$ years to discover such an important algorithm by Bruno Buchberger since the Greek figured out the proof of finding the greatest common divisor of two natural numbers, which is fully described in Euclid's *Elements*.

One of the main difficulties in solving a system of polynomial equations in more than one unknown is that the polynomial ring $k[X, Y]$ in two unknowns X and Y over a field k is a unique factorization domain (*UFD*) but it is not a principal ideal domain (*PID*) in general.

EXAMPLE 1.2. *We will later prove that $k[X, Y]$ is a UFD. But, the maximal ideal (X, Y) in $k[X, Y]$ is never generated by a single polynomial in $k[X, Y]$, because there is no non-constant polynomial in $k[X, Y]$ which* *divides both polynomials X and Y, irreducible over k. Thus, $k[X, Y]$ is not a PID. [Remark: We will introduce all the needed definitions and terminologies very soon.]*

Because of this, we can not uniquely determine quotients and a remainder of division of a polynomial $f(X, Y) \in \mathbb{Q}[X, Y]$ by non-zero polynomials $g_1(X, Y)$ and $g_2(X, Y)$:

$$f(X, Y) = a_1(X, Y)g_1(X, Y) + a_2(X, Y)g_2(X, Y) + r(X, Y).$$

For example, exactly because of this kind of difficulties, it is not easy to decide whether a given polynomial $f(X, Y)$ is a member of an ideal I that is generated by two polynomials $g_1(X, Y)$ and $g_2(X, Y)$. By definition, we know that

$$f(X, Y) \in (g_1(X, Y), g_2(X, Y)) \text{ if and only if } r(X, Y) \equiv 0.$$

Therefore, in order to computationally answer this fundamental belonging question we need to develop a computable means, that is, an algorithm, which can uniquely tell us whether the remainder $r(X, Y)$ in the above expression is the zero polynomial or not, in general. This problem is nowadays called **an ideal membership question**.

On the other hand, even a high school student knows how to divide out a polynomial $f(X) \in \mathbb{Q}[X]$ by a non-zero polynomial $g(X)$:

$$f(X) = a(X)g(X) + r(X)$$

in a unique manner. Thus, everybody knows whether a given polynomial $f(X)$ is a member of the ideal J generated by $g(X)$ or not by simply computing the remainder $r(X)$ using the division algorithm for polynomials in one variable.

Later in this book, we will apply The theory of Groebner basis to compute the discriminant of cubic or quartic equations, for instance.

8. Several Basic Definitions in Algebra

Before we go into an algorithmic aspects of Non-Linear Algebra, we are going to define several basic definitions in Algebra. In the long run, this will make the readers very comfortable in learning Algebra.

8.1. A First Circle.

DEFINITION 1.2. *A* ***field*** *F is an integral domain where every non-zero element a is invertible, i.e., for each $a \neq 0$ there exists b in F such that $ab = 1$.*

Such an a is called a **left inverse** of b and such a b is called a **right inverse** of a.

EXAMPLE 1.3. *$\mathbb{Q}$ = the set of all rational numbers, $\mathbb{R}$ = the set of all real numbers, and $\mathbb{C}$ = the set of all complex numbers.*

DEFINITION 1.3. *An* ***integral domain*** *D is a commutative ring where the* ***cancellation law*** *holds, i.e., if $ab = ac$ and $a \neq 0$ then $b = c$ in D.*

REMARK 1.2. *Thus, an integral domain is the place where you can cancel out a non-zero element, even without multiplying (mostly non-existing in D) its inverse on both sides of the equality under consideration.*

EXAMPLE 1.4. *Any fields, $\mathbb{Z}$ = the set of all integers, and $\mathbb{Q}[X]$ = the set of all polynomials in one unknown X with rational coefficients.*

DEFINITION 1.4. *A* ***commutative ring*** *K is a ring whose multiplication is commutative.*

EXAMPLE 1.5. *Any integral domains and the quotient ring $\mathbb{Z}/n\mathbb{Z}$ where n is a positive integer.*

DEFINITION 1.5. *A* ***ring*** *R is an Abelian group together with multiplication $R \times R \to R$ satisfying the following rules:*

(1) *(Existence of multiplicative identity) There exists 1 in R such that $r.1 = r = 1.r$ for any element r of R.*

(2) *(Associativity)* $a(bc) = (ab)c$.

(3) *(Distributive law) Multiplication is distributive over addition in both directions, i.e.,* $a(b+c) = ab + ac$ *and* $(a+b)c = ac + bc$.

EXAMPLE 1.6. *The set of all* $n \times n$ *matrices, where* $n = 1, 2, 3, ...$

REMARK 1.3. *Thus, there are two binary operations in a ring and they are usually called "addition (denoted by* $+$*)" and "multiplication (denoted by* $\cdot$*)".*

DEFINITION 1.6. *An* ***Abelian group*** A *is a group whose binary operation,* $+$*, is commutative, i.e.,* $a + b = b + a$.

REMARK 1.4. *Its "additive identity" is usually denoted by* 0.

DEFINITION 1.7. *A* ***group*** G *is a set together with a binary operation* $*$ *satisfying the following rules:*

(1) *(Existence of identity) There exists an element* e *in* G *such that* $g * e = g = e * g$ *for each element* g *of* G.

(2) *(Associativity)* $a * (b * c) = (a * b) * c$.

(3) *(Invertibility) Every element of* G *is invertible.*

EXAMPLE 1.7. $\mathbb{Z}$.

REMARK 1.5. *We "****know****" a* **set** S *only when we know each of its elements. When have a finite set, we have no problem at all. But, most of the sets we are going to deal with in mathematics are infinite sets. For example,*

$$\mathbb{N} = \text{the set of all natural numbers } 1, 2, 3, ...$$

*is the most important infinite set whose infiniteness we all know. We recall that the traditional meaning of English word "natural" is "God-given". Natural numbers are also called "****counting**** numbers" or "****ordering**** numbers" for an obvious reason that they are always used in counting and/or ordering objects.*

REMARK 1.6. *If you read the above definitions backward, you will be able to read them in a proper logical order in defining them.*

We need to define additional ones:

DEFINITION 1.8. *A* ***vector space*** *V over a field F is a module over the commutative ring F.*

EXAMPLE 1.8. *$\mathbb{R}^2$ = the real plane, and $\mathbb{R}^3$ = the Euclidean 3-space.*

DEFINITION 1.9. *A (left)* ***module*** *M over a commutative ring K is an Abelian group together with an operation (called scalar multiplication): $. : K \times M \to M$ satisfying the following rules: for each a, b in K and for each m, n in M,*

(1) $1.m = m$
(2) $(ab).m = a.(b.m)$
(3) $(a + b).m = a.m + b.m$
(4) $a.(m + n) = a.m + a.n$

EXAMPLE 1.9. *Any commutative ring K is a module over itself.*

EXAMPLE 1.10. *Any abelian group A is a module over $\mathbb{Z}$.*

Thus, there could be no binary operation on a set, we have one binary operation on a group G, and we have two binary operations $+$ and $\cdot$ on a ring R

DEFINITION 1.10. *An element a of a commutative ring R is a* ***zero divisor*** *if $a \neq 0$ and there exist $b \neq 0$ such that their product $ab = 0$.*

LEMMA 1.4. *A commutative ring R is an integral domain if and only if R has no zero divisors.*

PROOF. Suppose that R is an integral domain, that is, it satisfies the cancellation law. If $ab = 0$ and $a \neq 0$, then since

$$ab = 0 = a \cdot 0$$

holds, we cancel out a from both sides to conclude that $b = 0$,Thus, there are no non-zero b satisfying $ab = 0$ where $a \neq 0$. Conversely, it it has no zero divisors, it exactly means that $ab = 0$ implies either $a = 0$ or $b = 0$. Suppose that $ab = ac$ and $a \neq 0$. Then we have

$$a \cdot (b - c) = 0.$$

Then since $a \neq 0$ by hypothesis, $b - c$ must be zero, i.e., $b = c$, which is all about the cancellation law. □

DEFINITION 1.11. *A **binary operation** on a set S is a map, $S \times S \longrightarrow S$, from their Cartesian product $S \times S$ to S.*

We use maps to compare sets:

THEOREM 1.2. *Let $f\colon X \longrightarrow Y$ be a map.*

(1) *f is one-to-one, i.e., **injective**, if and only if it has a left inverse.*
(2) *f is onto, i.e., **surjective**, if and only if it has a right inverse.*

PROOF. Suppose that f is injective. For any y in the image of f we have only one preimage x, so we define this y to be sent back to the x. For any y in Y outside the image of f, we define this y to be sent to the same particular element x_0 of X of our choice. This defines a map $g\colon Y \longrightarrow X$ such that their composition $g \circ f$ is the identity map: $X \longrightarrow X$. This proves the $\Rightarrow$ direction of the first assertion. The converse of the first assertion is obvious.

Now suppose that f is surjective. Then we note that the domain X of the function f is partitioned into the disjoint union of various non-empty preimages of the elements $y \in Y$. Thus, by the Axiom of the choice function, which is equivalent to the Axiom of choice, we can choose an element out of each preimage of elements of Y under the map f, and this defines a function $g\colon Y \longrightarrow X$ such that $f \circ g$ is the identity map: $Y \longrightarrow Y$. This proves the $\Rightarrow$ direction of the second assertion. The converse of the second assertion is obvious. □

REMARK 1.7. *In particular, we note that our proof also tells us that such inverses are not uniquely determined.*

DEFINITION 1.12. *A map is called **bijective** if it is both injective and surjective.*

COROLLARY 1.1. *A map f is bijective if and only if it has both a right inverse and a left inverse. When this is the case, any right inverse of f is equal to any left-inverse of f.* □

COROLLARY 1.2. *A map f is bijective if and only if it has a (two-sided) inverse f^{-1}. When this is the case, any two inverses of f are the same, and*

$$(f^{-1})^{-1} = f. \qquad \square$$

DEFINITION 1.13. *A map ϕ from a module M to a module N over the same commutative ring K is called a* ***linear*** *map or linear transformation of modules if for each $m, m' \in M$ and for each $\alpha \in K$,*

$$\begin{aligned} \phi(m+m') &= \phi(m)+\phi(m'), \\ \phi(\alpha m) &= \alpha\phi(m) \end{aligned}$$

hold.

DEFINITION 1.14. *An* ***isomorphism*** *of two modules M and N is a linear map from M to N that is also one-to-one and onto.*

DEFINITION 1.15. *A map ϕ from a commutative ring R to another commutative ring S is called a* ***homomorphism*** *of rings if for each r and $r' \in R$,*

$$\begin{aligned} \phi(r+r') &= \phi(r)+\phi(r'), \\ \phi(rr') &= \phi(r)\phi(r'), \ \textit{and} \\ \phi(1) &= 1 \end{aligned}$$

hold.

DEFINITION 1.16. *An* ***isomorphism*** *of two commutative rings R and S is a homomorphism from R to S that is one-to-one and onto.*

The readers are strongly recommended to memorize the above definitions in the given order.

There is one theorem which is really essential to us.

THEOREM 1.3. *(**Mathematical Induction**) A set of natural numbers that contains the number 1 and the number $n+1$ whenever it contains a number n must be the set of all numbers.* □

DEFINITION 1.17. *A **subfield** F of a field k is a subring of k that is also a field.*

DEFINITION 1.18. *A **subring** S of a commutative ring R is a subset of R satisfying*

(1) *$1 \in S$, where 1 is the (only) multiplicative identity element of R.*
(2) *$a - b \in S$, whenever $a, b \in S$.*
(3) *$ab \in S$, whenever $a, b \in S$.*

DEFINITION 1.19. *A **subgroup** H of a group G is a subset of G satisfying*

(1) *$e \in H$, where e is the (only) identity element of G.*
(2) *$ab \in H$, whenever $a, b \in H$.*
(3) *$a^{-1} \in H$, whenever $a \in H$.*

DEFINITION 1.20. *An (left) **ideal** I of a commutative ring R is an Abelian subgroup of R under addition satisfying*

$$ra \in I,$$

whenever $a \in I$ and $r \in R$.

In other words, a subset I of a commutative ring R is an ideal if it satisfies the following conditions:

(1) $0 \in I$.
(2) If $a, b \in I$ then $a + b \in I$.
(3) If $a \in I$ and $r \in R$, then $ra \in I$.

8.2. A Second Circle.

REMARK 1.8. *From now on, every ring R is assumed to be an **integral domain**.*

REMARK 1.9. *In most cases, ideals I in a ring R are infinite sets. Since we can no longer directly deal with an infinite set any more, and since our set I has an algebraic structure, we naturally seek, as a next best treatment, to know whether an ideal I under consideration can be generated by a finite subset of I or not, and if it is the case, then we would like to compute its generators explicitly if ever possible.*

DEFINITION 1.21. *A **principal ideal domain PID** is a unique factorization domain (UFD) where every ideal is generated by a single element.*

DEFINITION 1.22. *A **unique factorization domain UFD** is an integral domain where every non-unit (i.e., non-invertible) element is expressed as a product of powers of irreducible elements in an essentially unique manner, i.e., up to associates.*

DEFINITION 1.23. *Let R be an integral domain.*

(1) *Two elements a, b in R are **associates** if one is a unit multiple of another, i.e., $a = ub$ for some unit, i.e., an invertible element, $u \in R$.*
(2) *A non-zero element in R is called **irreducible** if it is not a unit (i.e., invertible) and it has no proper divisor.*
(3) *An element a in R is a **proper divisor** of another element $b \in R$ if $b = aq$ for some $q \in R$ and neither a nor q is a unit (i.e. invertible).*
(4) *An element a in R is called **prime** if a is not zero and not a unit, and if a divides a product bc of two elements b, c of R then it divides one of the factors, b or c.*

Therefore, a prime element in an integral domain R is always irreducible (Lemma 2.3), but not conversely.

REMARK 1.10. *We also note that*

(1) *the definition of an irreducible element is exactly the same as that of a prime number, and*

(2) *the definition of a prime element is exactly the same as a well-known property of a prime number.*

EXAMPLE 1.11. *In the ring $\mathbb{Z}[\sqrt{-5}]$ of Gauss integers of integers, which is an integral domain consisting of the numbers of the form $a+b\sqrt{-5}$ with $a, b \in \mathbb{Z}$, we have two different factorizations:*

$$6 = 2 \cdot 3 = (1 + \sqrt{-5}) \cdot (1 - \sqrt{-5}).$$

This tells us that both 2 and 3 are irreducibles but not primes there.

LEMMA 1.5. *$\mathbb{Z}$ is a PID.*

PROOF. Let I be an ideal in the integral domain $\mathbb{Z}$. If $I = (0)$, we are done. Otherwise i.e., if $I \supset (0)$, then let d be the smallest non-zero positive number among non-zero elements of I. Then $(d) \subseteq I$. For the reverse inclusion, we apply the division algorithm by d to any non-zero element f of I:

$$f = ad + r$$

for uniquely determined quotient a and remainder r such that $0 \leq r < d$. Then since $r = f - ad \in I$, r must be equal to 0; consequently, $I \subseteq (d)$. □

Applying a similar argument, one can prove:

LEMMA 1.6. *The polynomial ring $k[x]$ in one variable x over a field k is a PID.*

PROOF. Since the zero ideal (0) is certainly generated by the element 0, we may assume that an ideal I of $k[x]$ contains a non-zero element d. If I contains a non-zero constant polynomial d in the bottom field k, then $1 = d^{-1}d \in I$; hence $I = (1) = k[x]$. Otherwise, we choose d be a non-zero non-constant polynomial in I of minimum degree. We claim that I is the principal ideal generated by d. Certainly, by definition, $(d) \subseteq I$. To prove the reverse direction, we go as follows: For any polynomial $f \in I$, we have

$$f = ad + r$$

for uniquely determined quotient polynomial a and the remainder polynomial r such that $0 \leq \deg(r) < \deg(d)$. Then since

$$r = f - ad \in I,$$

and since $\deg(r) < \deg(d)$, we must have $r \equiv 0$ (i.e., the zero polynomial) because of our choice of the polynomial d, having minimal degree among all of non-zero non-constant polynomials in I; thus $f = ad$, showing that $I \subseteq (d)$. □

On the other hand, thanks to **Gauss's Lemma** on primitive polynomials in $\mathbb{Z}[x]$, we have

THEOREM 1.4. *$\mathbb{Z}[x]$ is a UFD.* □

This will be proved in the next chapter (Theorem 2.7).

With some works in the field F of fractions of a *UFD* R but still under the same spirit of Gauss's, one can prove

THEOREM 1.5. *The polynomial ring $R[x]$ in one variable over a UFD R is again a UFD.* □

COROLLARY 1.3. *The polynomial ring $k[x, y]$ in two variables x, y over a field k is a UFD.*

PROOF. Since every *PID* is a *UFD*, $k[x]$ is a *UFD*. Then since $k[x, y]$ is $k[x][y]$, $k[x, y]$ is also a *UFD*. □

REMARK 1.11. *Since our study requires a good understanding of the so-called division process,we are going to review basic facts for a factorization process of expressing a bigger one in terms of smaller ones in the next chapter. Since these facts are usually treated purely axiomatically, we need to study how we may compute those objects that are both defined axiomatically and whose existence is proved axiomatically, if ever possible. As we shall see later, we must be far away from the application of the Axiom of Choice, or its equivalent ones, for instance, the Well-Ordering Principle, or the Zorn's Lemma, which is not a lemma but falsely called so because of a historical reason.*

9. A Concluding Remark

We may recall that the Well-Ordering Principle, which is equivalent to the Axiom of Choice in Mathematics, is in fact a simple generalization to any set X of a property P satisfied by the set $\mathbb{N}$ of all natural numbers.

The Well-Ordering Principle says that every set X has an ordering $>$ on it such that it satisfies the following property P: each non-empty subset S of X has a least element in S itself.

But, the main problem with this Principle is that the set of natural numbers $\mathbb{N}$ is the only known set equipped with an ordering $>$ on it satisfying such a property P, to our best knowledge until now, so that a "logical" generalization of the very particular property P of $\mathbb{N}$ to an arbitrary set X is quite difficult for us to reasonably accept it, although we nowadays accept it as an Axiom in Mathematics.

For instance, even on the set $\mathbb{Z}$ of all integers, we have so far failed to figure out such an order $>$.

Exercises for Chapter 1

EXERCISE 1.1. *Prove that to say that the cancellation law holds in a commutative ring R is the same as to say that R has no zero-divisors, i.e., If $ab = 0$ for $a, b \in R$ then one of either a or b is zero.*

EXERCISE 1.2. *Can you define a field consisting of only 2 elements?*

EXERCISE 1.3. *Can you define a field consisting of only 3 elements?*

EXERCISE 1.4. *Can you define a field consisting of only 4 elements?*

EXERCISE 1.5. *Can you see that every Abelian group is a $\mathbb{Z}$-module?*

EXERCISE 1.6. *Derive a formula for the sum:*

$$S_5(n) = 1^5 + 2^5 + ... + n^5,$$

and prove that your formula holds for each $n \in \mathbb{N}$. (Hint: Consider the difference $(k+1)^6 - k^6$ when k.runs through from 1 to n.)

EXERCISE 1.7. *Can you explain a few of the main differences between exact computation and numerical computation?*

EXERCISE 1.8. *When and why do you use software packages like MAPLE, REDUCE, which are nowadays called Computer Algebra System, or Matlab?*

EXERCISE 1.9. *For numerical computations, a few methods of parallel computation had been developed and standardized: OpenMP and MPI. Are there any standardized methods for parallel computation for exact computations?*

CHAPTER 2

Factorization

1. Introduction

In 1637, Fermat (1601 – 1665) stated that the Diophantine equation

$$x^n + y^n = z^n,$$

with integer $n > 2$ has no positive integer solutions x, y, z. For small numbers $n = 3$, or even $n \leq 125,000$ the statement has been proved. But trying to adapt the method of proof that works for prime $n = 3$ to the general case, Kummer (1810 – 1893) thought he found a proof, but Dirichlet (1805 – 1859) observed that at one point the argument Kummer gave had a gap. It later turned out that the difficulty consist in the fact the cyclotomic fields in which we are led to operate are not a unique factorization domain if prime $n > 19$. Until that time, most people thought that the uniqueness of factorization of natural numbers into primes is so obvious that a formal proof is virtually a waste of time. But, Kummer and Dirichlet recognized that in an only slightly more general setting the statement of essential uniqueness of factorizations is actually false. More historical accounts in this direction are, for instance, in E. Grosswald's book, *Topics from the Theory of Numbers*, 2nd Edition.

We first recall that multiplication of two numbers is a function as well as addition of two number is so. Thus, whoever computes them *in that direction* the result of such computations on two numbers respectively are the same.

But, *the reverse computation* in such a situation is not always very happy. Here is our first example:

Example 2.1. *In the ring* $\mathbb{Z}[\sqrt{-5}]$, *which is an extended integral domain of* $\mathbb{Z}$, *we have*

$$6 = 2 \cdot 3 = (1 + \sqrt{-5})(1 - \sqrt{-5}).$$

We also have

$$3^2 = 3 \cdot 3 = (2 + \sqrt{-5}) \cdot (2 - \sqrt{-5}).$$

This example tells us, first of all, that the number 6 has two different factorizations and hence the ring $\mathbb{Z}[\sqrt{-5}]$ is not a *UFD* (Unique Factorization Domain). We recall that $\mathbb{Z}$ is a *PID* (Principal Ideal Domain), and hence it is automatically a *UFD*. But, its extended ring $\mathbb{Z}[\sqrt{-5}]$ is not a *UFD*; hence it is not a *PID*.

In other words, this example says that, depending on where we are working, the "smaller ones", even though you have figured them out properly, whose multiplication-results in the given number 6, are *not* uniquely determined.

We note that the above example is not the only such example where the uniqueness of factorization fails. In fact, (although we are not going to prove it in this book) one can prove that

Theorem 2.1. *Let* d *be a negative integer. The ring* $\mathbb{Z}[\sqrt{d}]$ *of integers (in the imaginary quadratic field* $\mathbb{Q}[\sqrt{d}]$*) is a Unique Factorization Domain (UFD) if and only if* $d = -1, -2, -3, -7, -11, -19, -43, -67$, *or* -163. $\square$

Remark 2.1. *Here is some historical remark: Gauss proved the* $\Leftarrow$ *direction of the above theorem. He also conjectured that there were no others. Thus much more difficult part of the theorem above was finally proved by Baker and Stark in* 1966, *after the problem had been worked on for more than* 150 *years.*

On the other hand, the following example should still confuse at least some of you:

EXAMPLE 2.2. *The number* 5 *in* $\mathbb{N}$ *or* $\mathbb{Z}$ *is prime and hence it is irreducible, but if we enter the ring* $\mathbb{Z}[i]$ *of Gauss integers, which by definition, consists of those numbers of the form* $a+bi$*, where* $a, b \in \mathbb{Z}$*, we have apparently* 4 *different factorization of the number* 5*:*

$$5 = (2+i)(2-i) = (-2-i)(-2+i) = (1-2i)(1+2i) = (-1+2i)(-1-2i),$$

where $i = \sqrt{-1}$*, that is,* i *is a number whose square is equal to* -1*. Thus, the number* 5 *is not prime and hence not irreducible anymore there. Later, we will learn that, in a UFD, irreducible elements are prime elements, and conversely.*

On the other hand, we know that it is a UFD, for instance, by Theorem 2.1 above. But, since we have exactly 4 *units,* $\pm 1, \pm i$ *in* $\mathbb{Z}[i]$*, these apparently different factorizations*

$$(2+i)(2-i),\ (-2-i)(-2+i),\ (1-2i)(1+2i),\ (-1+2i)(-1-2i)$$

are essentially the same, i.e., equivalent. This is because the 4 *associates of* $2+i$ *are*

$$2+i,\ -2-i,\ -1+2i,\ 1-2i,$$

and the 4 *associates of* $2-i$ *are*

$$2-i,\ -2+i,\ 1+2i,\ -1-2i.$$

Therefore, we still have only one factorization of the number 5 *in ring* $\mathbb{Z}[i]$ *of Gauss integers.*

REMARK 2.2. *We recall that a field* k *is, by definition, an integral domain where every non-zero element is invertible. Thus, no meaningful factorization of an element of a field* k *is ever possible. From now on we assume that all our rings are commutative rings.*

We have seen that the uniqueness of factorization of numbers fails when we extend $\mathbb{Z}$ to a certain larger integral domain $\mathbb{Z}[\sqrt{d}]$.

This kind of problem was a main difficulty of Kummer when he thought that he found a proof of Fermat's Last Theorem. In 1983 Faltings proved that there are only finitely many integer solutions for Fermat's Last Theorem, for any given value of n. It is now known that it is finally (and completely) proved by an English mathematician, Andrew Wiles in his famous paper, *Modular elliptic curves and Fermat's last theorem,* Pages 443-551, Annals of Mathematics 141(3), May 1995. It thus took about 350 years for its complete proof and it is one of the major triumphs of mathematics of the 20th century.

We think this is sufficient enough for an introduction to this Chapter. In what follows, we will study factorizations in a more systematic or axiomatic manner.

Exercises for Chapter 2.1

EXERCISE 2.1. *Prove that* $\pm 1, \pm i$ *are the only units in* $\mathbb{Z}[i]$.

2. Factorization of Polynomials in One Variable

The main purpose of this chapter is a study of division in rings. Because it is modeled on properties of the ring $\mathbb{Z}$ of integers, we need to review these properties. The results in this particular section are a continuation of those in the second-half of the last chapter.

REMARK 2.3. *We note that as long as factorization of polynomials are concerned, treatment of those polynomials in one variable is completely different from treatment of those polynomials in more than one variable. This is mainly because a polynomial ring* $k[x]$ *in one variable over a field* k *is a UFD that is also a PID, whereas a polynomial ring* $k[x, y]$ *in two variables over a field* k *is a UFD but is not a PID.*

In this section, we will study polynomials in one variable only, because a proper treatment of polynomials in more than one variables does require the theory of Groebner bases, discovered and developed by Bruno Buchberger (1942 -) in late 20th *century.*

We start our discussion with an elementary school arithmetic stuff:

THEOREM 2.2. *(Division with Remainder) If a, b are two distinct natural numbers and $a \neq 0$, then there exist unique integers q, r such that*

$$b = aq + r, \tag{2.1}$$

with $0 \leq r < a$. □

Thus, whoever computes the quotient q and the remainder r, the results are going to be the same. Moreover, we have an **algorithm** for actually computing q and r for a given set of inputs: b and $a \neq 0$, which we learned it even when we are in an elementary school.

One immediate consequence of this theorem, which we already proved in the previous Chapter, is:

COROLLARY 2.1. *$\mathbb{Z}$ is a PID.* □

We note that given non-zero natural numbers a, b, the set defined by

$$a\mathbb{Z} + b\mathbb{Z} := \{ar + bs : r, s \in \mathbb{Z}\}$$

is again an ideal in $\mathbb{Z}$, the smallest ideal containing both principal ideals $a\mathbb{Z}$ and $b\mathbb{Z}$, and by this Corollary it is generated by a single element, say d:

$$d\mathbb{Z} = a\mathbb{Z} + b\mathbb{Z}.$$

The immediate question one can think of is: how to compute d explicitly.

LEMMA 2.1. $d = \gcd(a, b)$.

PROOF. Certainly, $d = ar + bs$ for some $r, s \in \mathbb{Z}$. Also, d clearly divides both a and b since $a = a \cdot 1 + b \cdot 0$ for instance. Now, if h divides both a and b, then a and b are elements of the ideal $h\mathbb{Z}$; hence $d\mathbb{Z} = a\mathbb{Z} + b\mathbb{Z} \subset h\mathbb{Z}$; hence h does divide d.This exactly means that d is the greatest common divisor of a and b. □

COROLLARY 2.2. $\gcd(a, b) = ar + bs$ *for some* $r, s \in \mathbb{Z}$. □

REMARK 2.4. *We note that the above treatments do not tell us how we compute* $\gcd(a, b)$, *r and s. To compute* $\gcd(a, b), r$, *and s, we apply the* *very well-known* ***division algorithm of Euclid*** *of linear time efficiency, which is known to us since about BC* 300*. Here is the division algorithm of Euclid for* $\gcd(a, b)$.

Algorithm (Euclid's algorithm for $\gcd(a, b)$) We may assume that $q_0 = b > a_1 = a \neq 0$. If b is a multiple of $a \neq 0$, then $a = \gcd(a, b)$. Otherwise, we continuously apply the above theorem to obtain:

$$\begin{aligned}
b &= q_1 a + r_1, \text{ with } r_1 > 0, \\
a &= q_2 a_1 + r_2, \text{ with } a_1 = r_1 \text{ and } r_2 > 0, \\
a_1 &= q_3 a_2 + r_3, \text{ with } a_2 = r_2 \text{ and } r_3 > 0, \\
&\quad \ldots\ldots\ldots \\
a_{k-2} &= q_k a_{k-1} + r_k, \text{ with } a_{k-1} = r_{k-1} \text{ and } r_k > 0, \\
a_{k-1} &= q_{k+1} a_k + 0, \text{ with } a_k = r_k > 0 \text{ but } r_{k+1} = 0.
\end{aligned}$$

Then $r_1 > r_2 > \ldots > r_k > 0$ and hence $\gcd(a, b) = \gcd(r_1, r_2) = \gcd(r_2, r_3) = \ldots = \gcd(r_k, 0) = r_k$. □

This algorithm does work since

$$\gcd(b, a) = \gcd(b - aq, a) = \gcd(r, a) = \gcd(a, r),$$

and by rewriting these equations backwards starting from $\gcd(a, b)$, we can find those integers r and s in $\gcd(a, b) = ar + bs$.

REMARK 2.5. *We note that these r and s are not uniquely determined.*

EXAMPLE 2.3. *We know by inspection that* $\gcd(6, 29) = 1$*. On the other hand,*

$$\begin{aligned}
29 &= 4 \cdot 6 + 5, \\
6 &= 1 \cdot 5 + 1, \\
5 &= 5 \cdot 1 + 0.
\end{aligned}$$

Thus, $1 = 6 - 1 \cdot 5 = 6 - 1 \cdot (29 - 4 \cdot 6) = 5 \cdot 6 + (-1) \cdot 29$*; hence* $r = 5$ *and* $s = -1$*. Obviously, these r and s are not determined uniquely:*

$1 = 5 \cdot 6 + \{(-29) \cdot 6 + 29 \cdot 6\} + (-1) \cdot 29 = (-24) \cdot 6 + 28 \cdot 29$, *where* $6 \cdot 29 = \operatorname{lcm}(6, 29)$, *etc.*

LEMMA 2.2. *Let p be a prime number and let a, b be numbers. If p divides their product ab, then p divides at least one of them.*

PROOF. Suppose that the prime p does not divide a. Then $\gcd(p, a) = 1$; hence by the above corollary we have

$$1 = pr + as$$

for some $r, s \in \mathbb{Z}$. Multiplying both sides by b we see that p does divide b. □

THEOREM 2.3. *(**Fundamental Theorem of Arithmetic**) Every non-zero integer a can be written as a product*

$$a = cp_1 \cdots p_k, \tag{2.2}$$

where $c = \pm 1$, the p_i are prime numbers, and $k \geq 0$. This expression is unique except for the ordering of the prime factors p_i.

PROOF. First, we prove that a prime factorization exists: To prove this, it is enough to consider the case that a is greater than 1. By induction on a, we may assume that the existence is proved for all positive integers $b < a$. Either a is prime, in which case the product has one factor, or there is a proper divisor $b \neq a$. Then $a = bb'$, where both b and b' are smaller than a, and hence by induction hypothesis they can be factored into primes. Setting their factorization side by side gives a factorization of a.

Second, we prove that the factorization is unique: For this we suppose that

$$\pm p_1 \cdots p_n = \pm q_1 \cdots q_m.$$

The signs should certainly agree. Then we apply the above lemma with $p = p_1$: Since p_1 divides the product $q_1 \cdots q_m$ it divides some q_i, say q_1. Since q_1 is also prime, we must have that $p_1 = q_1$. We cancel p_1 from both sides and we proceed by induction, noting that 1 is never equal to a product of primes. □

REMARK 2.6. *Although the above factorization is unique up to units $c = \pm 1$, it is usually very difficult to compute those prime factors p_i when a is large. Moreover, the above proof does not tell us how one can compute those prime factors p_i efficiently.*

As we mentioned earlier, the structure of the ring $\mathbb{Z}$ of integers is closely analogous to that of a polynomial ring $k[x]$ in one variable x over a field k. Whenever a property of one of these two rings is derived, we should try to find an analogous property of the other.

EXAMPLE 2.4. *We have already seen that both $\mathbb{Z}$ and $k[x]$ are a PID in the previous chapter.*

DEFINITION 2.1. *A polynomial $p(x)$ with coefficients in a field k is called **irreducible** over k if it is not constant and if its only divisors of lower degree in $k[x]$ are constants.*

This means that the only way that p can be written as a product of two polynomials is $p = cp_1$ where c is a constant, i.e., an element of k, and p_1 is a constant multiple of p. The irreducible polynomials are analogous to prime integers, i.e., prime numbers. Since our polynomials have their coefficients in the field k, it is customary to normalize them by factoring out their leading coefficients, i.e., the coefficient of the monomial of the polynomial with highest degree, so that they become *monic*.

The proof of the following theorem is similar to the proof of the analogous statements for the ring of integers.

THEOREM 2.4. *Let k be a field and let $k[x]$ denote the polynomial ring in one variable x over a field k.*

- **a:** *If two polynomials f, g have no common non-constant factor, then there are polynomials $r, s \in k[x]$ such that $fr + gs = 1$.*
- **b:** *If an irreducible polynomial $p \in k[x]$ divides a product fg, then p divides one of the factors, f or g.*

c: *Every non-zero polynomial $f \in k[x]$ can be written as a product*

$$f = cp_1 \cdots p_k,$$

where c is a non-zero constant, the p_i are monic irreducible polynomials in $k[x]$, and $k \geq 0$. The factorization is unique, except for the ordering of the terms. □

REMARK 2.7. *To compute the r and s in part a of the above theorem, we simply apply the algorithm of Euclid for* $\gcd(f, g)$*, which is exactly the same as the above Algorithm, except for having* $\deg(r_1) > \dots \deg(r_{k-1}) > 0$*, instead of having* $_1 > \dots > r_{k-1}$ *there.*

The constant c in the third part of this theorem is analogous to the factor ± 1 in (2.2). These are units in their respective rings. The unit factors are there because we *normalized* primes to be positive, and irreducible polynomials to be monic. We can allow negative primes or non-monic irreducible polynomials if we wish. The unit factor can then be absorbed, if $k > 0$. But, this complicates the statement of uniqueness slightly.

The next examples shows that factorizations of polynomials in one variable over either the field $\mathbb{C}$ of complex numbers or the field $\mathbb{R}$ of real number are fairly easy.

EXAMPLE 2.5. *Over the field $\mathbb{C}$ of complex numbers, thanks to Gauss' Fundamental Theorem of Algebra, every polynomial of positive degree has a complex root α and hence has a divisor of the form $x - \alpha$. Therefore, the irreducible polynomials over $\mathbb{C}$ are linear, and the irreducible factorization of a polynomial over $\mathbb{C}$ has the form*

$$f(x) = c(x - \alpha_1) \cdots (x - \alpha_n), \tag{2.3}$$

where α_i are the roots of $f(x)$, repeated as necessary. Thus, the uniqueness of this factorization is not surprising.

REMARK 2.8. *Over the field $\mathbb{R}$ of real numbers, there are only two classes of irreducible polynomials: linear polynomials and irreducible quadratic polynomials. We recall that a real quadratic polynomial $x^2 + bx + c$ is* *irreducible over* $\mathbb{R}$ *if and only if its discriminant* $b^2 - 4c$ *is negative, in which case it has a pair of complex conjugate roots.* The fact that every irreducible polynomial over the complex numbers is linear implies that no higher-degree polynomial is irreducible over the reals. *This goes as follows: Suppose that a polynomial* $f(x)$ *has real coefficients* a_i *and that* α *is a complex, non-real root of* $f(x)$. *Then the complex conjugate* $\bar{\alpha}$ *is different from* α *and is also a root. For, since it is a real polynomial, the its coefficients* a_i *satisfy* $a_i = \bar{a}_i$, *and hence*

$$f(\bar{\alpha}) = a_n\bar{\alpha}^n + ... + a_1\alpha + a_0 = \bar{a}_n\bar{\alpha}^n + ... + \bar{a}_1\alpha + \bar{a}_0 = \overline{f(\alpha)} = \bar{0} = 0.$$

Thus the quadratic polynomial $g(x) := (x-\alpha)(x-\bar{\alpha}) = x^2 - (\alpha + \bar{\alpha})x + \alpha\bar{\alpha}$ *has real coefficients* $-(\alpha + \bar{\alpha})$ *and* $\alpha\bar{\alpha}$, *and moreover, both of its linear factors appear on the right side of the complex factorization (2.3) of* $f(x)$. *Thus,* $g(x)$ *divides* $f(x)$. *This means that the factorization of* $f(x)$ *into irreducible real polynomials is obtained by grouping conjugate pairs in the complex factorization of* $f(x)$.

REMARK 2.9. *But, factorization of polynomials in one variable is more complicated for* polynomials with rational coefficients *than for real or complex coefficients, because there exist irreducible polynomials in* $\mathbb{Q}[x]$ *of arbitrary degree.*

EXAMPLE 2.6. *For any* $n > 0$, *the polynomial* $x^n - 3x^{n-1} + 3$ *is irreducible over* $\mathbb{Q}$, *by Eisenstein's Criterion for irreducibility of polynomials with rational coefficients. We will later study his criterion in detail. (See Proposition 2.11).*

Neither the form of the irreducible factorization nor its uniqueness is intuitively clear for rational polynomials in one variable.

REMARK 2.10. *Because of several hypotheses of Eisenstein's Criterion, it helps us to determine only whether rational polynomials special type are irreducible or not, as we see in this Example. That is, for general rational polynomials Eisenstein's Criterion is not useful at all. The next trial we are going to do is to study of factorizations in* $\mathbb{Z}[x]$, *which in fact provides the main idea for the proof of Eisenstein Criterion, Proposition 2.11.*

The main reason why we want to move into $\mathbb{Z}[x]$ *may be the following. In there, we have a natural ring homomorphism* $\mathbb{Z}[x] \longrightarrow \mathbb{Z}_p[x]$, *defined by*

$$f(x) = a_n x^n + \cdots + a_0 \longmapsto \bar{f}(x) = \bar{a}_n x^n + \cdots + \cdot \bar{a}_0,$$

where $\mathbb{Z}_p := \mathbb{Z}/p\mathbb{Z}$ *is a field for each prime number* p; *hence the cardinality of* $\mathbb{Z}_p$ *is* p, *i.e., finite; hence in* $\mathbb{Z}_p[x]$, *there are only finitely many polynomials of* the same degree. *Here,* $\bar{a}_i$ *denotes the residue class of* $a_i \bmod p$. *We call* $\bar{f}(x)$ *the* ***residue*** *of* $f(x) \bmod p$. *Clearly, this ring homomorphism is induced by the projection:* $\mathbb{Z} \longrightarrow \mathbb{Z}_p$ *and its kernel is* $p\mathbb{Z}[x]$.

This ring homomorphism sometimes makes our life quite easier because it can be used to determine whether a given primitive *polynomial* $f(x)$ *is an irreducible polynomial or not. (Cf. Definition 2.5). That is, if there exists a prime integer* p, *not dividing* a_n, *such that the residue* $\bar{f}(x) \bmod p$ *is irreducible, then so is* $f(x)$. *Therefore, for a decision for smaller degree primitive polynomials to be irreducible or not, this homomorphism is very useful, as we shall explicitly see it in the next section.*

But it turned out that a close examination of factorization of rational polynomials requires us to introduce more fine terminologies and more abstract concepts in order to analyze the problem more clearly. This is basically done in the next section. Then, thanks to Gauss, we can prove that $\mathbb{Z}[x]$ *is a UFD, and other important results for later usage. This is treated in the section, called Gauss's Lemma below.*

For future reference, we have the following

PROPOSITION 2.1. *A polynomial* $f(x)$ *of degree* n *with coefficients in an integral domain* R *has at most* n *roots in* R.

PROOF. We know that $\alpha \in R$ is a root of f if and only if $x - \alpha$ divides $f(x)$. If it is the case, we can write $f(x) = (x - \alpha)q(x)$, where $q(x)$ is a polynomial of degree $n - 1$. If β is another root of $f(x)$, then $f(\beta) =$

$(\beta - \alpha)q(\beta) = 0$. Since R is an integral domain, it has no zero divisors. Thus, one of the two elements $(\beta - \alpha), q(\beta)$ must be zero. Thus, either $\beta = \alpha$, or $q(\beta) = 0$. Then by induction hypothesis, we may assume that $q(x)$ has at most $n-1$ roots in R. Then there are at most n possibilities for β. □

REMARK 2.11. *The fact that R is an integral domain is crucial to Proposition 2.1, as the following example shows.*

EXAMPLE 2.7. *Let R be the quotient ring $\mathbb{Z}/8\mathbb{Z}$. Then in the polynomial ring $R[x]$, we have*

$$\begin{aligned} x^2 - 1 &= (x+1)(x-1) = (x+3)(x-3) \\ &= (x+5)(x-5) = (x+7)(x-7). \end{aligned}$$

Thus, the polynomial $x^2 - 1$ has four roots $\bmod 8$, *and its factorization into irreducible polynomials is not unique.*

Exercises for Chapter 2.2

EXERCISE 2.2. *Let a, b be positive integers whose sum is a prime p. Prove that their greatest common divisor is 1.*

EXERCISE 2.3. *Prove that if n is a positive integer which is not a square of an integer, then $\sqrt{n}$ is not a rational number. For instance, $\sqrt{2}$ is not a rational number.*

EXERCISE 2.4. *Compute the greatest common divisor of the following polynomials:*

$$x^3 - 6x^2 + x + 4,$$

$$x^5 - 6x + 1.$$

EXERCISE 2.5. *Prove that there are infinitely many prime integers.*

3. *UFD*'s and *PID*'s

In the previous section, we have seen that factorization of polynomials in one variable is analogous to factorization of integers. Now, it is natural to ask whether other rings can have such properties. Relatively few such rings exist, but the ring $\mathbb{Z}[i]$ of Gauss integers is one interesting example. Before we study the so-called Gauss' Lemma, which is definitely needed in the study of irreducibility of polynomials in $\mathbb{Z}[x]$, we need to introduce more terminologies for a better study of factorization.

From now on, we assume that the given ring R is an integral domain, so that the Cancellation Law is available.

REMARK 2.12. *In order to describe factorization properly, we need the following definitions. Since we are too much familiar with the ring $\mathbb{Z}$ of integers, we have a tendency of identifying an irreducible element with a prime element, but when we are in an extended integral domain of $\mathbb{Z}$ it is not always the case, as we have already seen in examples in the previous section.*

DEFINITION 2.2. *Let R be an integral domain.*

(1) *Let $a, b \in R$. We say that an element a **divides** another element b (sometimes abbreviated $a|b$) if $b = aq$ for some $q \in R$.*
(2) *An element a in R is a **proper divisor** of another element $b \in R$ if $b = aq$ for some $q \in R$ and neither a nor q is a unit (i.e. an invertible element).*
(3) *A non-zero element in R is called **irreducible** if it is not a unit (i.e., invertible) and it has no proper divisor.*
(4) *Two elements $a, a' \in R$ are called **associates** if each divides the other.*
(5) *An element a in R is called **prime** if a is not zero and not a unit, and if a divides a product bc of two elements b, c of R then it divides one of the factors, b or c.*

It is easily seen that a, a' are associates if and only if they differ by a unit factor, that is, $a' = ua$ for some unit $u \in R$.

We also note that by definition, the only divisors of an irreducible elements are units and its associates.

LEMMA 2.3. *A prime element in an integral domain R is irreducible.*

PROOF. (By contradiction) Let $a \in R$ be a prime element. If it were not an irreducible element, then $a = bc$ for some non-units, i.e., non-invertible elements, b and c in R. Then since a is prime and since we obviously have $a|a$, a divides one of its factors, b or c. Say a divides b, so $b = ad$ for some $d \in R$. Then $a = bc = adc$; hence $1 = dc$ by the Cancellation Law, and hence c is invertible. But, this is a contradiction to our choice of b and c. Therefore, the prime a must be irreducible. □

REMARK 2.13. *The converse of this lemma does not hold in general, unless R is a UFD (Unique Factorization Domain). (See Example 2.9 below).*

The concepts of divisor, unit, and associate can be interpreted in terms of the *principal ideals* generated by any one of these elements.

REMARK 2.14. *This approach is definitely needed since it provides us a means to compare which one is bigger than or smaller than the other. In $\mathbb{Z}$, we obviously have a total ordering on it so that any proper factor of a number has a "smaller" size.*

DEFINITION 2.3. *An ideal I in a ring R is called* **principal** *if it is generated by a single element:*

$$I = (a) := \{ra : r \in R\}.$$

Then one can easily check that

(1) u is a unit if and only if $(u) = (1)$.
(2) a and a' are associates if and only if $(a) = (a')$.
(3) a divides b if and only if $(a) \supseteq (b)$.
(4) a is a proper divisor of b if and only if $(1) \supset (a) \supset (b)$, where the inclusion $\supset$ is a proper inclusion.

Since we are looking for establishing a theorem that is analogous to the Fundamental Theorem of Arithmetic in an integral domain R (Theorem 2.3 above), we may divide the statement of the theorem into two parts. First, a given element a must be a product of irreducible elements, and second, this product must be essentially unique.

Since the second part usually causes us more trouble than the first part, we first consider the first part in detail. For this, we assume that our element a is neither zero nor a unit; otherwise, we have no hope of writing it as a product of irreducible elements. Then we attempt to factor a, proceeding as follows: If a is irreducible itself, we are done. If not, then a has a proper factor, so it decomposes in some way as a product, $a = a_1 b_1$, where neither a_1 nor b_1 is a unit. We continue factoring a_1 and b_1 if possible, and we hope that this procedure terminates; in other words, we hope that after a finite number of steps all the factors at our hands are irreducibles. The condition that this procedure always terminates has a neat description in terms of principal ideals:

PROPOSITION 2.2. *Let R be an integral domain. Then the following two statements are equivalent:*

- **a:** *For every non-zero element a of R which is not a unit, the process of factoring a terminates after finitely many steps and results in a factorization $a = b_1 b_2 \cdots b_k$ of a into irreducible elements of R.*
- **b:** *R does not contain an infinite increasing chain of principal ideals*

$$(a_1) \subset (a_2) \subset (a_3) \subset ...,$$

where the inclusion $\subset$ is proper.

PROOF. Suppose that R contains an infinite increasing sequence $(a_1) \subset (a_2) \subset (a_3) \subset \ldots$. Then $(a_n) \subset (1)$ for every n because $(a_n) \subset (a_{n+1}) \subseteq (1)$. Since $(a_{n-1}) \subset (a_n)$, a_n is a proper divisor of a_{n-1}, say $a_{n-1} = a_n b_n$, where both a_n and b_n are not units. This provides a non-terminating sequence of factorizations of a_1: $a_1 = a_2 b_2 = a_3 b_3 b_2 = a_4 b_4 b_3 b_2 = \ldots$.

Conversely, such a sequence of factorizations gives us an increasing chain of ideals. □

REMARK 2.15. *The second condition of this proposition is often called the* ***ascending chain condition*** *for principal ideals. We will say, by definition, that* ***existence of factorizations*** *holds in R if the equivalent conditions of the proposition are true.*

Here is an example in which existence of factorizations fails.

EXAMPLE 2.8. *Let $R = k[x_1, x_2, ...]$, where for $m = 1, 2, ...,$ x_{m+1} is the 2^m-th root of x_1. That is, R is obtained by adjoining all 2^m-th roots of the variable x_1 to the polynomial ring $k[x_1]$, with relations, $x_2^2 = x_1, x_3^2 = x_2$, and so on. Then we can factor the element x_1 indefinitely in this ring, and correspondingly there is an infinite chain $(x_1) \subset (x_2) \subset ...$ of principal ideals.*

It turns out that we need infinitely many generators for a ring to make an example such as the one given above, so we will rarely encounter such rings.

In practice, the second part of the Fundamental Theorem (Theorem 2.3) is the one which gives the most trouble. Factorization into irreducible elements will usually be possible, but it will not be unique.

Units in a ring complicate the statement of uniqueness. It is clear that unit factors should be disregarded, since there is no end to the possibility of adding unit factors in pairs uu^{-1}. For the same reason, *associate* factors should be considered equivalent.

The units in the ring $\mathbb{Z}$ of integers are ± 1, and in this ring it was natural to normalize irreducible element (primes) to be positive; similarly, we may normalize irreducible polynomials in one variable with coefficients in a field k by insisting that they be monic. But, we do not have a reasonable way to normalize elements of an arbitrary integral domain, so we will allow some ambiguity.

It is actually neater to work with *principal ideals* than with elements: Associates generate the same principal ideal. However, it is not too cumbersome to use elements here, and we will stay with them.

DEFINITION 2.4. *We will call an integral domain a* ***Unique Factorization Domain (UFD)*** *if it satisfies the following properties.*

(1) *Existence of factorizations is true for R. In other words, the process of factoring a non-zero element a which is not a unit terminates after finitely many steps and yields a factorization* $a = p_1 \cdots p_m$, *where each* p_i *is irreducible.*
(2) *The irreducible factorization of an element is unique in the following sense: If a is factored in two ways into irreducible elements, say* $a = p_1 \cdots p_m = q_1 \cdots q_n$, *then* $m = n$ *and with suitable ordering of the factors,* p_i *is an associate of* q_i *for each* i.

Thus in the statement of uniqueness, associate factorizations are considered **equivalent**.

Here is an example where uniqueness of factorization fails.

EXAMPLE 2.9. *Let* $R = \mathbb{Z}[\sqrt{-5}]$, *consisting of all complex numbers of the form,* $a + b\sqrt{-5}$, *where* $a, b \in \mathbb{Z}$. *Since this ring is a subring of the field* $\mathbb{C}$ *of complex numbers, this ring is an integral domain. In this ring, units are* ± 1, *and the integer* 6 *has two essentially different factorizations in R:*

$$6 = 2 \cdot 3 = (1 + \sqrt{-5})(1 - \sqrt{-5}).$$

Using the ***norm function*** $N\colon R \longrightarrow \mathbb{N} \cup \{0\}$ *defined by*

$$N(\alpha) = \alpha\bar{\alpha} \ (= |\alpha|^2) = a^2 + 5b^2,$$

where $\alpha = a + b\sqrt{-5}$ *with* $a, b \in \mathbb{Z}$ *and* $\bar{\alpha} := a - b\sqrt{-5}$, *we can show, for instance, that* $1 + \sqrt{-5}$ *is irreducible, and it goes as follows:*

By definition, this function is clearly multiplicative:

$$N(\alpha\beta) = N(\alpha)N(\beta),$$

and moreover $N(\alpha)$ *is a positive integer unless* $\alpha = 0$.

We have $N(1+\sqrt{-5}) = 6 = 2 \cdot 3$, *which is in particular a product of prime numbers* 2 *and* 3 *in* $\mathbb{N}$.

Now, if $1+\sqrt{-5}$ *were reducible, then the norm of its proper factor,* $a+b\sqrt{-5}$, *with* $a, b \in \mathbb{Z}$, *must be equal to either the prime number* 2 *or the prime number* 3. *But,* $N(a+b\sqrt{-5}) = a^2+5b^2$ *is always strictly bigger than* 3, *unless* $b = 0$. *Consequently, we must have* $b = 0$. *If* $b = 0$, *then its proper factor* a *must be either* $\pm\sqrt{2}$ *or* $\pm\sqrt{3}$, *but both of them are not in* R. *Therefore,* $1+\sqrt{-5}$ *is indeed irreducible.*

By a similar argument, one can show that other factors, $2, 3$, *and* $1-\sqrt{-5}$ *are all irreducible in* R.

Thus, the prime numbers 2 *and* 3 *are not primes in* R *any more.*

For units in R, *we can easily prove that* $a+b\sqrt{-5}$ *is a unit in* R *if and only if* $N(a+b\sqrt{-5})$ $(= a^2+5b^2) = 1$. *That is,* units are those elements in R lying on the unit circle in the complex plane, centered at the origin.

Then by a similar argument as we have done above, we easily see that the only units in R *are* ± 1.

Thus, both 2 *and* 3 *are not associates of* $1 \pm \sqrt{-5}$, *which shows that the two factorizations above are essentially different and hence that* $R = \mathbb{Z}[\sqrt{-5}]$ *is not a unique factorization domain.*

The crucial property of prime integers is that if a prime divides a product, it divides one of the factors, and this property is used as a defining property for an element in an integral domain to be prime (Definition 2.2.)

PROPOSITION 2.3. *Let* R *be an integral domain. Suppose that existence of factorizations holds in* R. *Then* R *is a Unique Factorization Domain if and only if every irreducible element is prime.*

The proof is a simple extension of the arguments used in Lemma 2.2 and Theorem 2.3:

PROOF. If every irreducible is prime, then R is indeed a *UFD*, since existence of factorizations in R is guaranteed by hypothesis. Conversely, suppose that R is a *UFD* and let $a \in R$ be irreducible. To prove that a is prime, we assume that $a|bc$ with $b, c \in R$. Then $bc = ad$ for some $d \in R$. Since R is a *UFD*, we break b, c, and d into their irreducible factors to have:

$$\alpha p_1 \cdots p_k \cdot q_1 \cdots q_l = a \cdot r_1 \cdots r_m,$$

where α is a unit and every other factors are all irreducible. Thus, either one of the p_i's or one of the q_i's must be an associate of the irreducible a; hence $a|b$ or $a|c$. $\square$

COROLLARY 2.3. *In a UFD, every irreducible element is prime.* $\square$

Its proof is given in the proof of Proposition 2.3.

REMARK 2.16. *It is very important to distinguish the two concepts of irreducible element and prime element. They are equivalent in Unique Factorization Domains, but most rings contain irreducible elements which are not prime.*

EXAMPLE 2.10. *In the ring* $R = \mathbb{Z}[\sqrt{-5}]$, *the element* 2 *has no proper factor, so it is irreducible. But it is not prime because, although it divides the product* $6 = (1 + \sqrt{-5})(1 - \sqrt{-5})$, *it does not divide either factor.*

REMARK 2.17. *Since irreducible elements in a UFD are prime, the phrase* ***irreducible factorization*** *and* ***prime factorization*** *are synonymous. We can use them interchangeably when we are working in a UFD, but not otherwise.*

There is a simple way of deciding whether an element a divides another element b in a *UFD*, in terms of their irreducible (or prime) factorizations.

PROPOSITION 2.4. *Let* R *be a UFD, and let* $a = p_1 \cdots p_r$, *and* $b = q_1 \cdots q_s$ *be given prime factorization of two elements of* R. *Then* a *divides* b *if and only if* $s \geq r$ *and with a suitable ordering of the factors* q_i *of* b, p_i *is an associate of* q_i *for each* $i = 1, ..., r$. $\square$

COROLLARY 2.4. *Let* R *be a UFD, and let* a, b *be elements of* R *which are not both zero. Then there exists a* ***greatest common divisor*** $d = \gcd(a, b)$ *of* a, b, *i.e., an element* d *in* R *satisfying the following properties:*

(1) *d divides both a and b.*
(2) *if an element f of R divides both a and b, then f divides d.* □

It follows immediately from the second condition that any two greatest common divisors of a and b are in fact associates.

REMARK 2.18. *However,* the greatest common divisor need not have the form $ar + bs$ *for some* $r, s \in R$, *unless we are working in a PID* R.

EXAMPLE 2.11. *In the next section, we will prove that* $\mathbb{Z}[x]$ *is a UFD. In this ring,* 2 *and* x *are irreducible elements, so that they are primes, and* $\gcd(2, x) = 1$. *But* 1 *is not a linear combination of these two elements with integer polynomial coefficients.*

We recall that every ideal in the ring $\mathbb{Z}$ of integers is principal, which is a very important property of $\mathbb{Z}$. We have also proved that in an integral domain, every prime element is irreducible (Lemma 2.3). In our approach of proving that a *PID* is a *UFD*, we prove the following proposition first:

PROPOSITION 2.5. *In a PID, every irreducible element is prime.*

PROOF. Let a be irreducible. Assume that $a|bc$ but a does not divide b. We consider the ideal (a, b) in R generated by a and b. Since we are working in a *PID*, this ideal is generated by a single element, say d of R, $(a, b) = (d)$. We also note that this ideal properly contains the ideal (a) since a does not divide b by hypothesis; hence d is not an associate of a. Now, $(a, b) = (d)$ also implies that $d|a$ and $d|b$. Then since a is irreducible, d, which is a non-associate of a, must be a unit. Thus, $(a, b) = (1)$, i.e., $1 = ar + bs$ for some r, s in the ring. Thus, $c = car + cbs$ is divisible by a, which is what we wanted to prove. □

THEOREM 2.5. *A Principal Ideal Domain (PID) is a UFD.*

PROOF. Let R be a *PID*. Then every irreducible element of R is prime by the above proposition. Thus, according to Proposition 2.3, we need only to prove the existence of factorizations for R. By Proposition 2.2, this is equivalent to showing that R contains no infinite increasing chain of principal ideals. We argue by contradiction. Suppose that $(a_1) \subset (a_2) \subset (a_3) \subset \ldots$ is such a chain. Then since the union I of these ideals is an ideal of R, it is generated by a single element, say b, $I = (b)$. Then, b is contained in one of the (a_n); hence $I \subset (a_N) \subset (a_{N+1}) \subset \ldots \subset I$ for some N. In other words, our ascending chain of principal ideals is stationary, which contradicts the assumption that $(a_n) \subset (a_{n+1})$ for each n, and this completes the proof. □

REMARK 2.19. *The converse of Theorem 2.5 is not true in general. The ring* $\mathbb{Z}[x]$ *of integer polynomials is a UFD, but it is not a PID.*

PROPOSITION 2.6. *Let R be a PID and let p be a non-zero element of R. Then the followings hold and moreover they are equivalent to each other:*

- **a:** *$R/(p)$ is a field if and only if p is irreducible.*
- **b:** *The maximal ideals in a PID are those principal ideals generated by irreducible elements.*

PROOF. Since an ideal M is maximal if and only if the quotient ring R/M is a field, the two parts (a) and (b) are equivalent. More precisely, the $(b) \Longrightarrow (a)$ direction is rather obvious, because $R/(p)$ is a field if and only if (p) is a maximal ideal, which is in turn equivalent by part (b) to the statement that p is irreducible. Now, for another direction $(a) \Longrightarrow (b)$, we suppose part (a). For an irreducible element p of R, $R/(p)$ is a field; hence (p) is a maximal ideal. Now let M be a maximal ideal of R. Since we are in a *PID*, it is generated by a single element, say p of R, $M = (p)$. Since $R/(p)$ is a field, p is irreducible by part (a).

Now, we are going to prove the equivalence in part (b). First, let M be a maximal ideal. Then since we are working in a *PID*, $M = (p)$ for some $p \in R$. If p has a proper divisor b, then $(p) \subset (b) \subset (1)$ properly, so that $M = (p)$ is not a maximal ideal; hence p must be irreducible. Secondly, for an irreducible p in R, we know that the only divisors of p are the units and the associates of p. Therefore, the only principal ideals containing (p) are (1) and (p), that is, there are no other proper ideals properly containing (p); hence (p) is a maximal ideal. □

Exercises for Chapter 2.3

EXERCISE 2.6. *Prove that if R is an integral domain, then two elements a, b are associates if and only if they differ by a unit factor.*

EXERCISE 2.7. *Prove that $(a, b)[a, b] = ab$ in a principal ideal domain.*

EXERCISE 2.8. *Prove that $\mathbb{Z}[i]$ is a PID. [Hint: It is a Euclidean domain.]*

EXERCISE 2.9. *Prove that the following rings are Euclidean domains:*

(1) $\mathbb{Z}[\omega]$, *where* $\omega = e^{2\pi i/3}$.
(2) $\mathbb{Z}[\sqrt{-2}]$.

4. Gauss's Lemma

We first note that Theorem 2.4 allows us to fairly freely manipulate polynomials in one variable over a field k. This is mainly because $k[x]$ is a *PID*. In most of this section we fix our bottom field k to be the field $\mathbb{Q}$ of rational numbers, until we begin to mention that $\mathbb{Q}$ is the field of fractions of a *UFD* $\mathbb{Z}$.

Among rational polynomials in one variable are integral polynomials, and the set $\mathbb{Z}[x]$ of all integral polynomials in one variable, whose elements are main objects of our study in this section, is certainly a subring of $\mathbb{Q}[x]$; hence it is also an integral domain.

In this section, we are going to prove that $\mathbb{Z}[x]$ is a *UFD*, and its proof definitely needs the so-called Gauss's Lemma for primitive polynomials in $\mathbb{Z}[x]$.

EXAMPLE 2.12. *We note that $\mathbb{Z}[x]$ is not a PID, because the ideal $(2, x)$ in $\mathbb{Z}[x]$ generated by the prime number 2, which is necessarily an irreducible element of $\mathbb{Z}[x]$, and the irreducible polynomial x, which is obviously an irreducible element of $\mathbb{Z}[x]$, is never generated by a single element in $\mathbb{Z}[x]$.*

Then the theory developed to prove that $Z[x]$ is a *UFD* easily leads us to establish the following important result: If R is a *UFD*, then so is $R[x]$.

Here is our beginning example of an irreducible factorization in $\mathbb{Z}[x]$:

EXAMPLE 2.13. $6x^3 + 9x^2 + 9x + 3 = 3(2x + 1)(x^2 + x + 1) \in \mathbb{Z}[x]$.

REMARK 2.20. *We do not mention such a factorization as a "prime" factorization until we prove that $\mathbb{Z}[x]$ is a UFD.*

As we see from this example, irreducible factorizations are slightly more complicated in $\mathbb{Z}[x]$ than in $\mathbb{Q}[x]$. First, the prime integers are necessarily irreducible elements of $\mathbb{Z}[x]$, so they may appear in an irreducible factorization of an integral polynomial, i.e., a polynomial in $\mathbb{Z}[x]$. But in $\mathbb{Q}[x]$ irreducible elements are irreducible polynomials, and vice versa. Second, the irreducible factor $2x + 1$ is not a monic polynomial. If we want to stay with integer coefficients, we can not ask for monic factors.

DEFINITION 2.5. *An integral polynomial $f(x)$ is called* ***primitive*** *if its coefficients $a_0, ..., a_n$ have no common integer factor except for the units ± 1 and if its highest coefficient a_n is positive.*

Thus, irreducibility question of integral polynomials $f(x)$ in one variable is basically reduced to irreducibility question of its associated primitive polynomials, because the "whole integer factor" of $f(x)$ is easily determined by computing the greatest common divisor of coefficients of $f(x)$. At any rate, the integer factors of a polynomial $f(x) = a_n x^n + \cdots + a_0$ in $\mathbb{Z}[x]$ are common divisors of its coefficients $a_0, ..., a_n$ of $f(x)$.

LEMMA 2.4. *Every non-constant irreducible integral polynomial with positive leading coefficient is necessarily primitive, but not conversely.* □

The converse of this lemma requires some additional condition, as is described in Proposition 2.8 below.

LEMMA 2.5.

(1) *Every non-zero rational polynomial* $f(x) \in \mathbb{Q}[x]$ *can be written as a product*
$$f(x) = cf_0(x),$$
where c *is a rational number and* $f_0(x)$ *is a primitive integral polynomial in* $\mathbb{Z}[x]$.

(2) *Moreover, this expression for* f *is unique.*

(3) *The polynomial* $f(x)$ *has integer coefficients if and only if* c *is an integer.*

(4) *If (3) holds, then* $|c|$ *is the greatest common divisor of the coefficients of* $f(x)$, *and the sign of* c *is the sign of the leading coefficient of* $f(x)$.

DEFINITION 2.6. *The* rational number c *which appears in this lemma is called the* **content** *of* $f(x)$.

Thus, we observe that if $f(x)$ has integer coefficients, then the content divides $f(x)$ in $\mathbb{Z}[x]$. We also see that $f(x)$ is primitive if and only if its content is 1.

PROOF. To find $f_0(x)$, we first multiply $f(x)$ by an integer to clear the denominators in its coefficients. This will give us a polynomial $f_1(x)$ with integer coefficients. Then we factor out the greatest common divisor of the coefficients of $f_1(x)$ and adjust the sign of the leading coefficient. The resulting polynomial $f_0(x)$ is primitive, and $f = cf_0$ for some rational number c. This proves existence.

To prove uniqueness, suppose that $cf_0(x) = dg_0(x)$, where $c, d \in \mathbb{Q}$ and $f_0(x)$ and $g_0(x)$ are primitive polynomials. We will show that $c = d$ and $f_0(x) = g_0(x)$. Clearing denominators of both c and d reduces us to the case that c and d are integers. Let $\{a_i\}, \{b_i\}$ denote the coefficients of $f_0(x)$ and $g_0(x)$, respectively. Then $ca_i = db_i$ for each i. Since the greatest common divisor of $\{a_0, ..., a_n\}$ is 1, c is the greatest common divisor of $\{ca_0, ..., ca_n\}$. Similarly, d is a greatest common divisor of $\{db_0, ..., db_n\} = \{ca_0, ..., ca_n\}$. Thus, $c = \pm d$ and $f_0(x) = \pm g_0(x)$. Then since the primitive polynomials $f_0(x)$ and $g_0(x)$ have positive leading coefficients, we conclude that $f_0(x) = g_0(x)$, and $c = d$.

Now, if $f(x)$ has integer coefficients, clearing of the denominators is not necessary; hence c is an integer, and up to sign it is the greatest common divisor of the coefficients, as stated. □

Now, we are ready to prove Gauss's Lemma:

THEOREM 2.6. *(Gauss's Lemma) A product of two primitive polynomials in* $\mathbb{Z}[x]$ *is again primitive.*

PROOF. Let the primitive polynomials be denoted by f and g, and let $h = fg$ be their product. Since the leading coefficients of f and g are positive, the leading coefficient of h is also positive. To prove that h is primitive, that is, to prove that no prime integer p divides all the coefficients of h, we consider the ring homomorphism $\mathbb{Z}[x] \longrightarrow \mathbb{Z}_p[x]$ for a prime integer p, which is introduced in Remark 2.10. Since both f and g are primitive, no prime integer p will divide their integer coefficients, and hence both residues $\bar{f}$ and $\bar{g}$ are not the zero polynomial in $\mathbb{Z}_p[x]$ for each prime integer p. Then since $\mathbb{Z}_p[x]$ is an integral domain, their product $\bar{f}\bar{g}$, which is exactly the same as $\bar{h} = \overline{fg}$, is also a non-zero polynomial in $\mathbb{Z}_p[x]$ for each prime integer p. This exactly means that $h \notin p\mathbb{Z}[x]$, the kernel of the ring homomorphism, for each prime integer p. Thus, h is primitive. □

REMARK 2.21. *The following proposition relates divisibility between two integral polynomials* $f(x)$ *and* $g(x)$ *in* $\mathbb{Z}[x]$ *with divisibility between the same* $f(x)$ *and* $g(x)$ *in* $\mathbb{Q}[x]$.

More precisely, we note that a division in an integral domain is always a division in its field of fractions, so our movement in this direction gives no difficulty to us. On the other hand, if coefficients of polynomials are all invertible, i.e., units, then it is quite easy to handle coefficients of polynomials. Thus, we have a tendency of preferring to work on divisions over a field. But, in order to properly relate a division over a field of fractions of a UFD R *to a division in its subring* R, *where our objects for* *division are assumed to be in the subring R, we definitely need Gauss's Lemma, as the proof of the following Proposition demonstrates.*

The main reason why we prefer to work on a UFD is to keep a good track on each irreducible factor of the involved objects while we are in division process.

We also recall that every field has only two ideals, (0) and (1), so it is a PID; hence it is a UFD.

PROPOSITION 2.7.

a: *Let f, g be rational polynomials in $\mathbb{Q}[x]$, and let f_0, g_0 be their associated primitive polynomials in $\mathbb{Z}[x]$. If f divides g in $\mathbb{Q}[x]$, then f_0 divides g_0 in $\mathbb{Z}[x]$.*

b: *Let f be a primitive polynomial in $\mathbb{Z}[x]$, and let g be any integral polynomial in $\mathbb{Z}[x]$. Suppose that f divides g in $\mathbb{Q}[x]$, say $g = fq$, with a rational polynomial q in $\mathbb{Q}[x]$. Then q is an integral polynomial in $\mathbb{Z}[x]$, and hence f divides g in $\mathbb{Z}[x]$.*

c: *Let f, g be integral polynomials in $\mathbb{Z}[x]$. If they have a common non-constant rational polynomial factor in $\mathbb{Q}[x]$, then they have a common non-constant integral polynomial factor in $\mathbb{Z}[x]$ too, and conversely.*

PROOF. The converse of each statement is trivial.

To prove part *a*, we may clear denominators so that f and g become primitive. Then part *a* is a consequence of part *b*. More precisely, according to Lemma 2.5 we write $f = cf_0$, $g = dg_0$, where c and d are associated contents in $\mathbb{Q}$, respectively, and f_0, g_0 are associated primitives, respectively. Thus, if f divides g in $\mathbb{Q}[x]$, then we have $dg_0 = \alpha cf_0$, with some $\alpha \in \mathbb{Q}[x]$, so that $g_0 = d^{-1}\alpha cf_0$. Thus, the primitive f_0 divides the integral polynomial g_0 in $\mathbb{Q}[x]$. Thus, part *a* is a consequence of part *b*.

To prove part *b*, we apply Lemma 2.5 in order to write the quotient in the form $q = cq_0$, where q_0 is primitive and $c \in \mathbb{Q}$. Then by Gauss's Lemma (Theorem 2.6), fq_0 is primitive, and the equation $g = cfq_0$ shows that fq_0 is the primitive polynomial g_0 associated to g. Therefore $g = cg_0$

is the unique expression for g referred to in Lemma 2.5, and the rational number c is the content of g. Then since g is an integral polynomial by hypothesis, it follows that c is in fact an integer, and hence q is also an integral polynomial.

Finally, to prove part c, suppose that f, g have a common rational polynomial factor h in $\mathbb{Q}[x]$. We may assume that h is primitive, and then by part b, h divides both f and g in $\mathbb{Z}[x]$. $\square$

COROLLARY 2.5. *If a non-constant integral polynomial f is irreducible in $\mathbb{Z}[x]$, then it is also irreducible in $\mathbb{Q}[x]$.*

PROOF. This is true by part b of the proposition above. $\square$

PROPOSITION 2.8. *Let f be an integral polynomial with positive leading coefficient. Then f is irreducible in $\mathbb{Z}[x]$ if and only if either*

(1) *f is a prime integer, or*
(2) *f is a primitive polynomial which is irreducible in $\mathbb{Q}[x]$.*

PROOF. Suppose that f is irreducible. As in Lemma 2.5, we may write $f = cf_0$, where f_0 is primitive. Since f irreducible, this can not be a proper factorization; hence either c or f_0 is 1. If $f_0 = 1$, then f is constant, and to be irreducible, a constant polynomial must be a prime integer. Otherwise, i.e., if $c = 1$, then f is primitive, and it is irreducible in $\mathbb{Q}[x]$ by Corollary 2.5.

The converse, that integer primes and primitive irreducible polynomials are irreducible element of $\mathbb{Z}[x]$, is clear. $\square$

EXAMPLE 2.14. *The polynomial $3(2x + 1)$ is irreducible over $\mathbb{Q}$ but is not irreducible over $\mathbb{Z}$.*

Let $f(x)$ be an integral polynomial in $\mathbb{Z}[x]$. Then it is also in $\mathbb{Q}[x]$. Then since $\mathbb{Q}[x]$ is a *PID*, it is a *UFD*; hence we have a unique factorization of $f(x)$ in $\mathbb{Q}[x]$ up to a unit:

$$f(x) = cq_1(x) \cdots q_n(x),$$

where c is a unit and each $q_i(x)$ are rational irreducible monic polynomials in $\mathbb{Q}[x]$.

In order to derive a factorization of $f(x)$ in $\mathbb{Z}[x]$ from this factorization, we simply apply the following procedure: First, we clear out the denominator of $q_1(x)$ and take out the greatest common divisor of all integer coefficients of the resulting integral polynomial in order to have in $\mathbb{Q}[x]$

$$f(x) = c_1 q_1'(x) q_2(x) \cdots q_n(x),$$

where c_1 is another unit in $\mathbb{Q}$, $q_1'(x)$ is primitive polynomial in $\mathbb{Z}[x]$ which is necessarily irreducible there, and $q_2(x) \cdots q_n(x)$ is still a rational polynomial in $\mathbb{Q}[x]$. This says that the integral polynomial $f(x)$ is divided by a primitive polynomial $q_1'(x)$ in $\mathbb{Q}[x]$. Therefore, by part b of Proposition 2.7 the product $c_1 \cdot q_2(x) \cdots q_n(x)$ is in fact an integral polynomial in $\mathbb{Z}[x]$. But since this is not yet good enough, we then apply the same treatment to $q_2(x)$ as we have done to $q_1(x)$ to have

$$f(x) = c_2 q_1'(x) q_2'(x) q_3(x) \cdots q_n(x),$$

where c_2 is another unit in $\mathbb{Q}$, $q_2'(x)$ is primitive polynomial in $\mathbb{Z}[x]$ which is necessarily irreducible there, and $q_3(x) \cdots q_n(x)$ is a rational polynomial in $\mathbb{Q}[x]$. If we repeat this "conversion" process at most n times, we will end up with

$$f(x) = c_n q_1'(x) \cdots q_n'(x),$$

where c_n is a unit in $\mathbb{Q}$, each $q_i'(x)$ primitive irreducible polynomial in $\mathbb{Z}[x]$; hence the product $q_1'(x) \cdots q_n'(x)$ is again primitive by Gauss's Lemma (Theorem 2.6). Then, again by part b of Proposition 2.7 whose proof also requires Gauss's Lemma, c_n is in fact an integer! In this way, we have derived a factorization of $f(x)$ in $\mathbb{Z}[x]$ from its always-existing unique factorization in the *UFD* $\mathbb{Q}[x]$. *In other words, we have proved that existence of factorization holds in* $\mathbb{Z}[x]$.

Having this process at our hands, we are on our way in proving that $\mathbb{Z}[x]$ is a *UFD* using Proposition 2.3.

PROPOSITION 2.9. *Every irreducible element of $\mathbb{Z}[x]$ is a prime element.*

PROOF. Let f be irreducible, and suppose that f divides in $\mathbb{Z}[x]$ a product gh, where $g, h \in \mathbb{Z}[x]$. By Proposition 2.8 we have only two cases.

Case 1. $f = p$ is a prime integer. We write $g = cg_0$ and $h = dh_0$ with integers c, d, as in Lemma 2.5. Then g_0h_0 is again primitive by Gauss's Lemma, and hence some coefficient a of the expansion of the product g_0h_0 must not be divisible by p. On the other hand, since p divides the product gh, p must divides all the coefficients of the expansion of the product gh; hence in particular p must divide the integer product acd. Then since p does not divide a, it must divide either c or d; hence p divides either g or h.

Case 2. f is a primitive polynomial which is irreducible in $\mathbb{Q}[x]$. In this case, f is a prime element of $\mathbb{Q}[x]$ since $\mathbb{Q}[x]$ is a *PID*. Thus, f divides either g or h in $\mathbb{Q}[x]$. Then by part b of Proposition 2.7, f divides either g or h in $\mathbb{Z}[x]$. □

THEOREM 2.7.

(1) *The polynomial ring $\mathbb{Z}[x]$ is a Unique Factorization Domain (UFD).*
(2) *Every non-zero integral polynomial $f(x)$ in $\mathbb{Z}[x]$ which is not ± 1 can be written as a product*

$$f(x) = \pm p_1 \cdots p_m \cdot q_1(x) \cdots q_n(x),$$

where the p_i are prime integers and the $q_i(x)$ are irreducible primitive polynomials in $\mathbb{Z}[x]$.
(3) *The expression is unique up to arrangement of the factors.*

PROOF. We have already seen existence of factorizations in $\mathbb{Z}[x]$ above. Thus, by Proposition 2.9 and Proposition 2.3 we completes our proof. □

Now let R be a *UFD*, and let F be its field of fractions. Then $R[x]$ is a subring of $F[x]$, and the results of this section can be copied, replacing $\mathbb{Z}$ with R and $\mathbb{Q}$ with F throughout. The only change to be made is that instead of normalizing primitive polynomials it is better to allow ambiguity caused by unit factors, as in the previous section. The main results are these:

THEOREM 2.8. *Let R be a UFD with field of fractions F.*

a: *Let f, g be polynomials in $F[x]$, and let f_0, g_0 be their associated primitive polynomials in $R[x]$. If f divides g in $F[x]$, then f_0 divides g_0 in $R[x]$.*

b: *Let f be a primitive polynomial in $R[x]$, and let g be any polynomial in $R[x]$. Suppose that f divides g in $F[x]$, say $g = fq$, with a polynomial q in $F[x]$. Then q is a polynomial in $R[x]$, and hence f divides g in $R[x]$.*

c: *Let f, g be polynomials in $R[x]$. If they have a common non-constant factor in $F[x]$, then they have a common non-constant factor in $R[x]$ too, and conversely.*

d: *If a non-constant polynomial f is irreducible in $R[x]$, then it is irreducible in $F[x]$, and conversely.*

e: *$R[x]$ is a UFD.*

The proof of Theorem 2.8 follows the pattern established for the ring $\mathbb{Z}[x]$, and we omit it. □

Since $R[x_1, ..., x_n]$ is isomorphic to $R[x_1, ..., x_{n-1}][x_n]$, we obtain the next corollary:

COROLLARY 2.6. *The polynomial rings $\mathbb{Z}[x_1, ..., x_n]$ and $F[x_1, ..., x_n]$, where F is a field, are UFD's.* □

Thus, the ring $\mathbb{C}[x, y]$ of complex polynomials in two variables is a *UFD*. In contrast to the case of one variable, however, where every complex polynomial is a product of linear ones, complex polynomials in two variables are often irreducible, and hence prime.

Exercises for Chapter 2.4

EXERCISE 2.10. *Let a, b be elements of a field F, with $a \neq 0$. Prove that a polynomial $f(x)$ over F is irreducible over F if and only if $f(ax+b)$ is irreducible over F.*

EXERCISE 2.11. *Let $F = \mathbb{C}(x)$, and let $f(x,y), g(x,y) \in \mathbb{C}[x,y]$. Prove that if $f(x,y)$ and $g(x,y)$ have a common factor in $F[y]$, then they also have a common vector in $\mathbb{C}[x,y]$.*

EXERCISE 2.12. *Prove that two integer polynomials are relatively prime in $\mathbb{Q}[x]$ if and only if the ideal they generate in $\mathbb{Z}[x]$ contains an integer.*

5. Explicit Factorization of Polynomials

We first recall that over the field $\mathbb{C}$ of complex numbers, thanks to Gauss' Fundamental Theorem of Algebra, every polynomial of positive degree has a complex root α and hence has a divisor of the form $x - \alpha$. Therefore, the irreducible polynomials over $\mathbb{C}$ are linear, and the irreducible factorization of a polynomial over $\mathbb{C}$ has the form

$$g(x) = c(x - \alpha_1) \cdots (x - \alpha_n), \tag{2.4}$$

where α_i are the roots of $g(x)$, repeated as necessary. Thus, the uniqueness of this factorization is not surprising.

And, over the field $\mathbb{R}$ of real numbers, there are only two classes of irreducible polynomials: linear polynomials and irreducible quadratic polynomials. We recall that a real quadratic polynomial $x^2 + bx + c$ is irreducible over $\mathbb{R}$ if and only if its discriminant $b^2 - 4c$ is negative, in which case it has a pair of complex conjugate roots.

What we want are the irreducible factors of a rational polynomial $g(x)$ in $\mathbb{Q}[x]$. By clearing out the denominator of $g(x)$ and pulling out the gcd of the resulting polynomial, we can write

$$g(x) = cg_0(x),$$

where $g_0(x)$ is a primitive integral polynomial in $\mathbb{Z}[x]$, which is what Lemma 2.5 all about, and hence by Corollary 2.5 this amounts to determining the irreducible factors in $\mathbb{Z}[x]$. Thus, in this section we want to seek explicit methods of determining the factors of a given integral polynomial

$$f(x) = a_n x^n + \cdots + a_1 x + a_0. \tag{2.5}$$

Linear factors of $f(x)$ can be found fairly easily: If $b_1x + b_0$ divides $f(x)$, then b_1 should divide a_n and b_0 should divide a_0. Then since there are only finitely many integers which divide a_n and a_0, we can try all possibilities. In each case, we carry out the division and determine whether the remainder is zero or not. Or, we may substitute the rational number $r = -b_0/b_1$ into $f(x)$ to see if it is a root.

REMARK 2.22. *Though things are not so clear for factors of higher degree, it is known that showed that the factors can be determined with a finite number of computations. His method is based on the Lagrange interpolations formula, but unfortunately this method requires too many steps to be practical except for factors of low degree, and a lot of work has been done on the problem of efficient computations.*

We we have mentioned before, one of the most useful methods is then computation modulo p, using the homomorphisms $\mathbb{Z}[x] \longrightarrow \mathbb{Z}_p[x]$ defined by

$$f(x) = a_n x^n + \cdots + a_0 \longmapsto \bar{f}(x) = \bar{a}_n x^n + \cdots + \cdot\bar{a}_0,$$

for various prime integers p, where $\mathbb{Z}_p := \mathbb{Z}/p\mathbb{Z}$ is a field for each prime number p. If our polynomial $f(x)$ factors in $\mathbb{Z}[x]$, so is, its residue $\bmod p$ under this map for each prime integer p, i.e., its image in $\mathbb{Z}_p[x]$ for each prime integer p. Then since there are only finitely many polynomials of the same degree in $\mathbb{Z}_p[x]$, all factorizations there can be carried out in finitely many steps.

PROPOSITION 2.10. *Let $f(x) = a_n x^n + \cdots + a_1 x + a_0 \in \mathbb{Z}[x]$ be an integral polynomial, and let p be a prime integer which does not divide the leading coefficient a_n. If the residue $\bar{f}(x)$ of $f(x)$ $\bmod p$ in $\mathbb{Z}_p[x]$ is irreducible there, then $f(x)$ is irreducible in $\mathbb{Q}[x]$, but not conversely.*

PROOF. This follows from an inspection of the homomorphism $\mathbb{Z}[x] \longrightarrow \mathbb{Z}_p[x]$. We need the assumption that p does not divide a_n in order to rule out the possibility that a factor $g(x)$ of $f(x)$ could reduce to a constant in $\mathbb{Z}_p[x]$. This assumption is preserved if we replace $f(x)$ with its associated primitive polynomial. Thus we may assume that $f(x)$ is primitive. Since p does not divide the leading coefficient a_n, the degrees of $f(x)$ and $\bar{f}(x)$ are the same. Now, if $f(x)$ factors in $\mathbb{Q}[x]$, by Corollary 2.5 it also factors in $\mathbb{Z}[x]$. Let $f(x) = g(x)h(x)$ be a proper factorization in $\mathbb{Z}[x]$. Since $f(x)$ is primitive, $g(x)$ and $h(x)$ have positive degree. Since $\deg(f(x)) = \deg(\bar{f}(x))$ and $\bar{f}(x) = \bar{g}(x)\bar{h}(x)$, it follows that $\deg(g(x)) = \deg(\bar{g}(x))$ and $\deg(h(x)) = \deg(\bar{h}(x))$; hence $\bar{f}(x) = \bar{g}(x)\bar{h}(x)$ is a proper factorization in $\mathbb{Z}_p[x]$, which shows that $\bar{f}(x)$ is reducible there. Then by Corollary 2.5 again, we complete our proof. □

The converse of this proposition does not hold in general.

EXAMPLE 2.15. *One easily checks by trial and error that the polynomial* $f(x) = x^4 - 10x^2 + 1$ *is irreducible over* $\mathbb{Z}$. *But,* $\bar{f}(x) \bmod p$ *for* $p = 2, 3, 5, 7$ *at least, is reducible over* $\mathbb{Z}_p$*:*

$$\begin{aligned}
f(x) \bmod 2 &= x^4 + 1 = (x^2 + 1)^2, \\
f(x) \bmod 3 &= x^4 + 2x^2 + 1 = (x^2 + 1)^2, \\
f(x) \bmod 5 &= x^4 + 1 = (x^2 + 2)(x^2 + 3), \\
f(x) \bmod 7 &= x^4 + 4x^2 + 1 = (x^2 + 6x + 6)(x^2 + x + 6).
\end{aligned}$$

Can you show that $\bar{f}(x) \bmod p$ *is in fact irreducible over* $\mathbb{Z}_p$ *for some prime* p*?*

Thus, the method of reduction modulo p will not always work. But it does work quite often:

Suppose that we suspect that a given integral polynomial $f(x) \in \mathbb{Z}[x]$ is irreducible. Then we can try reduction modulo p for a few low primes, $p = 2$ or 3 for instance, and hope that $\bar{f}(x)$ turns out to be of the same degree and irreducible. If so, we will have proved that $f(x)$ is indeed irreducible over $\mathbb{Z}$. Since $\mathbb{Z}_p$ is a field, $\mathbb{Z}_p[x]$ is a *PID*; hence Theorem 2.4 still holds for $\mathbb{Z}_p[x]$.

Before we go further, we want to explain a method, called the **sieve of Eratosthenes**, of determining the prime integers p less than a given number n. The method goes as follows: We list the numbers from 2 to n. We know that 2 is prime, then we cross out all multiples of 2 from our list. Then 3 is prime, since it is not divisible by any smaller prime. Then we cross out all multiples of 3 from our list, and we keep going on until we reach up to n. Then the left-overs are prime numbers less than n.

This "**sieve method**" will also determine the irreducible polynomials in $\mathbb{Z}_p[x]$. We list all polynomials, degree by degree, and the cross out products.

EXAMPLE 2.16. *The linear polynomials in $\mathbb{Z}_2[x]$ are x and $x+1$, and there are irreducible there. The polynomials of degree 2 are x^2, x^2+1, x^2+x, and x^2+x+1. Then the first three are divisible by x or $x+1$, so the last one is the only irreducible polynomial of degree 2 in $\mathbb{Z}_2[x]$. We keep going on for polynomials of degree 3, and so on. Here is a list of irreducible polynomials of degree ≤ 4 over $\mathbb{Z}_2$:*

$$\begin{aligned}
&x, x+1;\\
&x^2+x+1;\\
&x^3+x^2+1, x^3+x+1;\\
&x^4+x^3+1, x^4+x+1, x^4+x^3+x^2+x+1.
\end{aligned}$$

By trying the polynomials in the above list, we can factor all polynomials of degree $9 = (4+3+2+1)-1$ or less in $\mathbb{Z}_2[x]$, and hence we can extend this list to another list of irreducible polynomials of degree ≤ 9 in $\mathbb{Z}_2[x]$, etc. and etc.

EXAMPLE 2.17. *Thus, $x^4-6x^3+12x^2-3x+9$ is irreducible in $\mathbb{Q}[x]$; and hence in $\mathbb{Z}[x]$ because it is primitive. For, its residue in $\mathbb{Z}_2[x]$ is x^4+x+1 and it is irreducible there. We note that we can not conclude that it is* *irreducible over $\mathbb{Q}$ from Eisenstein's Criterion in the next Proposition, because the constant term is divisible by 3^2.*

EXAMPLE 2.18. *The monic irreducible polynomials over degree 2 over $\mathbb{Z}_3$ are the followings:*

$$x^2+1, x^2+x-1, x^2-x-1.$$

We also note that reduction modulo p may help describe the factorization of a polynomial even though the residue is reducible:

EXAMPLE 2.19. *Consider the polynomial $f(x) = x^3+6x+3$ for example. Reducing modulo 3, we obtain x^3 and this does not look like quite promising. However, suppose that $f(x)$ were reducible, say $(ax+b)(cx^2+dx+e) = x^3+6x+3$. Then the residue of $ax+b$ would have to divide x^3 in $\mathbb{Z}_3[x]$, which would imply $b \equiv 0 \bmod 3$. Similarly, we could conclude $e \equiv 0 \bmod 3$. But it is impossible to satisfy both of these conditions, because $be = 3$. Therefore, no such factorization exists, and $f(x)$ is irreducible over $\mathbb{Q}$.*

The principle at work in this Example is called the **Eisenstein Criterion**.

PROPOSITION 2.11. *(Eisenstein Criterion) Let $f(x) = a_nx^n + \cdots + a_0 \in \mathbb{Z}[x]$ be an integral polynomial, and let p be a prime integer. Suppose that the coefficients of $f(x)$ satisfies the following conditions:*

(1) p does not divide the leading coefficient a_n;
(2) p divides all other coefficients $a_{n-1}, \ldots, a_0$;
(3) p^2 does not divide the constant term a_0.

Then f is irreducible in $\mathbb{Q}[x]$. If moreover f is primitive, then it is irreducible in $\mathbb{Z}[x]$.

PROOF. Let $\bar{f}(x)$ denote the reside $\bmod\, p$ of $f(x)$. The first two hypotheses imply that $\bar{f}(x) = \bar{a}_nx^n$ and that $\bar{a}_n \neq 0 \bmod p$. If $f(x)$ were reducible in $\mathbb{Q}[x]$, then it would factor in $\mathbb{Z}[x]$ into factors of positive degree, say $f(x) = g(x)h(x)$ by Corollary 2.5. Then $\bar{g}(x)$ and $\bar{h}(x)$ would divide $\bar{a}_n x^n$, and hence each of these polynomials would be a monomial. Therefore all coefficients of $g(x)$ and $h(x)$, except the leading terms, would be divisible by p. But, let the constant coefficients of $g(x), h(x)$ be b_0, c_0. Then the constant coefficient of $f(x)$ is obviously $a_0 = b_0 c_0$. Since p divides both b_0 and c_0, p^2 divides a_0, which contradicts the third hypothesis. Therefore, $f(x)$ must be irreducible in $\mathbb{Q}[x]$. The last assertion follows from Proposition 2.8. □

One of the most important applications of the Eisenstein Criterion is to prove the irreducibility of the cyclotomic polynomial

$$x^{p-1} + x^{p-2} + \cdots + x + 1$$

is irreducible in $\mathbb{Q}[x]$, whose roots are the p-th root of unity, the powers of $\varsigma = e^{2\pi i/p}$:

COROLLARY 2.7. *Let p be a prime. The polynomial $f(x) = x^{p-1} + x^{p-2} + \cdots x + 1$ is irreducible in $\mathbb{Q}[x]$.*

PROOF. We first know that $(x-1)f(x) = x^p - 1$. If we make a change of coordinate $x = y + 1$, then it become

$$\begin{aligned} yf(y+1) &= (y+1)^p - 1 \\ &= y^p + \binom{p}{1} y^{p-1} + \binom{p}{2} y^{p-2} + \cdots + \binom{p}{p-1} y, \end{aligned}$$

where $\binom{p}{i} = p(p-1)\cdots(p-i-1)/i! \in \mathbb{N}$ is the i-th binomial coefficient. Thus, by dividing out both side by y, we obtain

$$f(y+1) = y^{p-1} + \binom{p}{1} y^{p-2} + \binom{p}{2} y^{p-3} + \cdots + \binom{p}{p-1}.$$

Now, for any $i < p$, p never divides each $i - j$ for $j = 0, 1, ..., i$; hence p can not divide $i!$. Thus, $\binom{p}{i}$ is divisible by p for $i < p$. Therefore, by Eisenstein Criterion $f(y+1)$ is irreducible in $\mathbb{Q}[y]$. Thus, $f(x)$ must be irreducible over $\mathbb{Q}$. □

Exercises for Chapter 2.5

EXERCISE 2.13. *Show that the following polynomials are irreducible over* $\mathbb{Q}$.

(1) $x^2 + 27x + 213$.
(2) $8x^3 - 6x + 1$.
(3) $x^5 - 3x^4 + 3$.

EXERCISE 2.14. *Factor* $x^5 + 5x + 5$ *into irreducible factors in* $\mathbb{Q}[x]$ *and in* $\mathbb{F}_2[x]$.

EXERCISE 2.15. *Factor* $x^3 + x + 1$ *in* $\mathbb{F}_p[x]$, *when* $p = 2, 3, 5$.

EXERCISE 2.16. *Factor* $x^4 + x^2 + 1$ *into irreducible factors in* $\mathbb{Q}[x]$.

EXERCISE 2.17. *Let* p *be a prime integer. Prove that the polynomial* $x^n - p$ *is irreducible over* $\mathbb{Q}$.

EXERCISE 2.18. *Prove that the polynomial* $x^2 + y^2 - 1$ *is irreducible over* $\mathbb{Q}$ *by the method of undetermined coefficients, that is, by studying the equation*

$$(ax + by + c)(a'x + b'y + c') = x^2 + y^2 - 1,$$

where a, b, c, a', b', c' *are unknown.*

6. Additive Factorization Theory?

Before we close the current section, we try to introduce very briefly the subject, called, the additive number theory, in which we study, for instance, the number $p(n)$ of different way of adding smaller numbers which is equal to a given natural number n. For example, we have in total 7 different ways of expressing the number 5:

$$\begin{aligned}&5,\\&4+1,\\&3+2,\\&3+1+1,\\&2+2+1,\\&2+1+1+1,\\&1+1+1+1+1.\end{aligned}$$

Here, we consider, for instance, $4 + 1$ is the "same" as $1 + 4$. Thus, $p(5) = 7$. The function $p(n)$ is called the "partition function", and many interesting questions on the property of $p(n)$ are answered, basically starting with the works of Euler, and then MacMahon, Hardy, Ramanujan, and Rademacher, etc. For example, Euler proved

THEOREM 2.9. *(Euler) Given a natural number n, one has*

$$p(n) = p(n-1)+p(n-2)-p(n-5)-p(n-7)+\cdots+(-1)^{j+1}p(n-n_j), \tag{2.6}$$

where $n_j = j(3j \pm 1)/2$ are the so-called pentagon numbers.

It is well-known that P.A. MacMahon used this recurrence formula in 1916 to compute the values of $p(n)$ for n up to $n = 200$, which is obtained from the famous identity of Euler:

THEOREM 2.10. *(Euler's pentagonal number theorem)*

$$\prod_{j=1}^{\infty}(1-t^j) = 1 + \sum_{j=1}^{\infty}(-1)^j t^{j(3j-1)/2}(1+t^j). \tag{2.7}$$

The proof of these two theorems are nowadays a standard material in most books in Number Theory.

According to George Andrew's account in p. 150 of his book, The Theory of Partitions, G.H. Hardy and S. Ramanujan were the first mathematicians who gave an asymptotic formula for $p(n)$ by power series in

1917, the first term of which is

$$\frac{1}{2\pi\sqrt{2}}\frac{d}{dn}\left(\frac{\exp(\frac{2\pi}{\sqrt{6}}\sqrt{n-\frac{1}{24}}}{\sqrt{n-\frac{1}{24}}}\right).$$

The first five terms of the Hardy-Ramanujan formula give the "correct" value of $p(200)$, which is $3,972,999,029,388$.

In 1937, H. Rademacher discovered an exact formula for $p(n)$ by a convergent series, which yields Hardy and Ramanujan's asymptotic formula as a corollary.

THEOREM 2.11. *(Rademacher) Given a natural number n, one has*

$$p(n) = \frac{1}{\pi\sqrt{2}}\sum_{k=1}^{\infty} A_k(n)\sqrt{k}\frac{d}{dn}\sqrt{\frac{\sinh((\pi/k)(2/3)(n-\frac{1}{24}))}{n-\frac{1}{24}}}, \tag{2.8}$$

where

$$A_k(n) = \sum_{\substack{h \bmod k \\ (h,k)=1}} \varpi_{h,k}\exp(-2\pi i n h/k),$$

where

$$\varpi_{h,k} = \exp(\pi i s(h,k)),$$

is a $24k$-th root of unity and where

$$s(h,k) = \sum_{\mu=1}^{k-1}\left(\frac{\mu}{k} - \left[\frac{\mu}{k}\right] - \frac{1}{2}\right)\left(\frac{h\mu}{k} - \left[\frac{hk}{k}\right] - \frac{1}{2}\right)$$

is the so-called Dedekind sum.

It is well-known that this theorem is based on the theory of modular forms and the proof of this hard-to-understand theorem is well-presented, for instance, in pp. 71-81 of George Andrew's book, The Theory of Partitions. This is the formula mathematicians up to now insisted to call an exact formula for $p(n)$, although it is very analytic in nature.

On the other hand, since the size of $p(n)$ increases very rapidly, for instance,

$$\begin{aligned} p(10) &= 42, \\ p(100) &= 190569292, \\ p(1000) &= 24061467864032622473692149727991, \end{aligned}$$

it requires a good computational means, for example, CAS(Computer Algebra System) like MAPLE or REDUCE for "Symbolic Computation", to compute the values $p(n)$ for various relatively large n, even if these theoretical results are available to us.

Exercises for Chapter 2.6

EXERCISE 2.19. *Using Theorem 2.10, prove Theorem 2.9.*

CHAPTER 3

Generators and Linear Relations

1. Modules and Determinants of R-Matrices

1.1. Introduction. When we study Mathematics we have seen many examples of "linear relations". For example, a set $\{v_1, ..., v_n\}$ of vectors is called *linearly independent* if

$$c_1 v_1 + ... + c_n v_n = 0,$$

for some scalars c_i's at least one of which is non-zero. Or, systems of linear equations, etc.

On the other hand, we recall that a vector space V over a field F is defined as a module over F. Then since every vector space, regardless of being finitely generated or not, has a basis, it is a free module of finite rank or infinite rank. (Remark: An official definition of a free module will be given in the next section.) As we shall see in the next section, this, namely, "every module over a field is free", is very special in the sense that modules M over a commutative ring R are not free in general.

In this chapter, we will systematically develop a theory for generators and linear relations among them up to a degree sufficient enough for us to study $F[t]$-modules, which will eventually lead us to better understand endomorphisms, i.e., linear operators, on a vector space in general. This will require us to study the structure of finitely generated abelian groups. We will need some knowledge on the polynomial ring $F[t]$, for instance, it is a *PID*, but it is going to be OK, because we are very familiar with the ring $\mathbb{Z}$ of integers.

We can also think of generators and non-linear relations among them, but they are better understood if we study them after learning the Theory of Groebner Bases first.

1.2. Modules. We first recall that a **module** M over a commutative ring R, or shortly, a R-module, is, by definition, an abelian group together with a scalar multiplication $R \times M \longrightarrow M$, sending $(r, v) \mapsto rv$, satisfying the following 4 axioms: for each $v, v' \in M$ and each scalar $r, s \in R$,

(1) $1v = v$,
(2) $(rs)v = r(sv)$,
(3) $(r + s)v = rv + sv$,
(4) $r(v + v') = rv + rv'$.

EXAMPLE 3.1. *If F is a field then an F-module is just a vector space over F.*

EXAMPLE 3.2. *Any commutative ring R is a module over itself. In this case, the above 4 axioms for the scalar multiplication for a module are simply those axioms defining it as a ring.*

The concepts of basis and linear independence can be carried over from vector spaces over a field K to modules over a commutative ring R without any change:

DEFINITION 3.1. *Let M be an R-module.*

a: *A set $\{v_1, ..., v_n\}$ of elements of M is said to* ***generate*** *(or* ***span****) M if each $v \in M$ is an R-linear combination:*

$$v = r_1v_1 + ... + r_nv_n,$$

with $r_i \in R$. In that case, the v_i are called ***generators****.*

b: *M is said to be* ***finitely generated*** *if it has a finite set of generators.*

c: *We call a set of elements $\{v_1, ..., v_n\}$ of M (linearly)* ***independent*** *if the following condition holds:*

$$r_1v_1 + ... + r_nv_n = 0 \text{ with } r_i \in R \qquad \text{implies} \qquad \text{each } r_i = 0 \text{ for } i = 1, ..., n$$

d: *Moreover, such a set is called a* ***basis*** *of M if it is a generating set of M.*

We then have an easy

LEMMA 3.1. *A subset $\{v_1, ..., v_n\}$ of an R-module M is a basis if and only if every element v of M is expressed as a linear combination*

$$v = r_1 v_1 + ... + r_n v_n$$

of the v_i's in a unique way, i.e., with unique coefficients r_i's in R. □

DEFINITION 3.2. (1) *A **homomorphism**, or **linear map**, of R-modules φ: $M \longrightarrow N$ is a map which is compatible with the laws of composition:*

$$\varphi(v + w) = \varphi(v) + \varphi(w) \qquad \text{and} \qquad \varphi(rv) = r\varphi(v),$$

for each $r \in R$ and for each $v, w \in M$.

(2) *A bijective homomorphism of R-modules is called an **isomorphism**.*

DEFINITION 3.3. *A **submodule** N of a module M is a non-empty subset of M such that it itself is closed under addition and scalar multiplication on M.*

Like in the case of linear maps of vector spaces, we can easily prove that

LEMMA 3.2. *The kernel of a homomorphism φ: $M \longrightarrow N$ of R-modules is a submodule of M, and the image of φ: $M \longrightarrow N$ is a submodule of N.* □

Now, the most non-trivial but obvious examples are the modules R^n of R-vectors.

DEFINITION 3.4. *The row or column vectors in R^n with entries in the ring R are called R-**vectors**. The laws of composition for R-vectors are* *the same as for vectors with entries in a field F:*

$$\begin{bmatrix} a_1 \\ \cdot \\ \cdot \\ \cdot \\ a_n \end{bmatrix} + \begin{bmatrix} b_1 \\ \cdot \\ \cdot \\ \cdot \\ b_n \end{bmatrix} = \begin{bmatrix} a_1 + b_1 \\ \cdot \\ \cdot \\ \cdot \\ a_n + b_n \end{bmatrix} \quad \textit{and} \quad r \begin{bmatrix} a_1 \\ \cdot \\ \cdot \\ \cdot \\ a_n \end{bmatrix} = \begin{bmatrix} ra_1 \\ \cdot \\ \cdot \\ \cdot \\ ra_n \end{bmatrix}.$$

Certainly, R^n together with such laws of composition is an R-module. This module is very special because it is a set of elements

$$e_i = \begin{bmatrix} 0 \\ \cdot \\ 1 \\ \cdot \\ 0 \end{bmatrix},$$

where the i-th entry is 1 but all other entries are 0, for each $i = 0, ..., n$. They are called the n **unit elements** of R^n, or the (**standard**) **basis** of R^n. As the standard basis in the case of the vector space F^n where F is a field, every element $v = (a_1, ..., a_n)^t \in R^n$ is uniquely expressed as an R-linear combination:

$$v = a_1 e_1 + ... + a_n e_n$$

of the n unit vectors in R^n with a_i as coefficients. The R-module R^n with the standard basis $\{e_i\}$ satisfies a very special property:

THEOREM 3.1. *Given n elements $v_1, ..., v_n$ of an R-module M, the linear map t: $R^n \longrightarrow M$ defined by*

$$\sum_{i=1}^{n} a_i e_i \longmapsto \sum_{i=1}^{n} a_i v_i$$

is the only linear map from R^n to M that sends e_i to v_i for each $i = 1, ..., n$.

PROOF. One can easily check that sending any $v = \sum_{i=1}^{n} a_i e_i \longmapsto \sum_{i=1}^{n} a_i v_i$ is a linear map from R^n to M. Of course this map sends each e_i to v_i by definition. If T is a linear map: $R^n \longrightarrow M$ which sends each

e_i to v_i for each $i = 1, ..., n$, then it certainly sends linear combinations to linear combinations: $\sum_{i=1}^{n} a_i e_i \longmapsto \sum_{i=1}^{n} a_i v_i$; hence $t = T$. □

Thus, R^n is a free R-module, according to the following definition.

DEFINITION 3.5. *Let X be a subset of an R-module F and $j\colon X \hookrightarrow F$ be the insertion of X in F. Then F is called* ***a free module on*** *X, or* ***free*** *on X,* ***(and X a set of free generators*** *for the module F) if we have the following universally repelling property: whenever we have a map $f\colon X \longrightarrow M$ to an R-module M, there exists a unique (repelling) linear map $t\colon F \longrightarrow M$ such that $f = t \circ j$. In this case, any set X of free generators is also called a* ***basis*** *for the free module F.*

Thus, the above theorem says

THEOREM 3.2. *The R-module R^n is a free module on the set $\{e_1, ..., e_n\}$ of the standard basis of R^n.* □

COROLLARY 3.1. *An R-module with a basis $\{v_1, ..., v_n\}$ is isomorphic to R^n.* □

DEFINITION 3.6. *Such an R-module is also called a* ***free*** *R-module.*

REMARK 3.1. *Thus, we immediately see that any surjective linear map onto a finitely generated R-module is completely determined by the images of free generators of a free R-module, and this observation is going to be the starting point of the main issue of this chapter, generators and linear relations among them. See one of the next section titles.*

We are now going to see that there are plenty of R-modules which are not free.

LEMMA 3.3. *Abelian groups and $\mathbb{Z}$-modules are equivalent concepts.*

PROOF. By definition, a $\mathbb{Z}$-module is an abelian group. Conversely, any abelian group V can be made into a $\mathbb{Z}$-module in exactly one way by the rules:

$$\begin{aligned} nv &= v + ... + v = \text{“}n \text{ times } v\text{”}, \\ (-n)v &= -(nv), \end{aligned}$$

for any positive integer n. Then those 4 axioms for a module is automatically satisfied. $\square$

COROLLARY 3.2. *The ring $\mathbb{Z}$ of integers is a $\mathbb{Z}$-module.*

EXAMPLE 3.3. *Since the set $\mathbb{Z}$ consists of infinitely many elements, any finite abelian group cannot be possibly isomorphic to a free $\mathbb{Z}$-module $\mathbb{Z}^n$ for some $n > 0$, unless it is the zero group; hence they are not free at all. A free $\mathbb{Z}$-module is also called a* ***free abelian group****.*

This example and Remark 3.1 certainly suggests that we need to extend the concept of quotient group to modules.

DEFINITION 3.7. *Let M be an R-module and N be its submodule. Then the set M/N of cosets $v + N$ for various $v \in M$, sometimes called the* ***quotient*** *of M by N, becomes an R-module under the following binary operations and scalar multiplication:*

$$\begin{aligned} (v+N)+(w+N) &= vw+N, \\ s.(v+N) &= sv+N, \end{aligned}$$

for each $v \in M$ and for each $s \in R$.

The best example one can easily understand quotient modules may be the following example:

EXAMPLE 3.4. *Let $6\colon\ \mathbb{Z} \longrightarrow \mathbb{Z}$ be "multiplication by 6". Then this map is clearly a homomorphism of $\mathbb{Z}$-modules, and hence its kernel $6\mathbb{Z}$ is a submodule of $\mathbb{Z}$. Thus, elements of the quotient module $\mathbb{Z}/6\mathbb{Z}$ are those remainders of integers v modulo 6. More precisely, by Euclid's algorithm we have a unique expression,*

$$v = 6q + r$$

with uniquely determined quotient q and the remainder r, $0 \leq r < 6$; by definition elements of $\mathbb{Z}/6\mathbb{Z}$ are $r+ 6\mathbb{Z}$.

We note that any two integers s, t with the same remainder r defines the same coset, $r + 6\mathbb{Z}$; that is, $s + 6\mathbb{Z} = t + 6\mathbb{Z}$, i.e., they, s and t, *"represents" the same coset $r + 6\mathbb{Z}$. (They are called* ***representatives*** *of the coset they represent.) We also note in this case that 6 divides $(s - t)$ if and only if $(s - t) + 6\mathbb{Z} = 0 + 6\mathbb{Z}$, i.e., $s - t \in 6\mathbb{Z}$.*

Following ideas described in this example, one can easily prove

PROPOSITION 3.1. *Then we have:*

(1) *The two rules in the above definition are* ***well-defined****. That is, they are independent of representatives of cosets they represent; and consequently M/N under these rules become an R-module, called the quotient module of M by N.*
(2) *The canonical map $\pi\colon M \longrightarrow M/N$ which sends $v \mapsto \bar{v} := v + N$ is a homomorphism of R-modules which is surjective with kernel N.*
(3) *If $f\colon M \longrightarrow M'$ is any homomorphism of R-modules whose kernel contains N, then there exists a unique homomorphism*

$$\bar{f} : M/N \longrightarrow M'$$

such that the involved diagram is commutative, that is, $f = \bar{f} \circ \pi$.
(4) *Moreover, if kernel of f is the same as N, then $\bar{f}$ gives an isomorphism of M/N with its image, $f(M)$.*
(5) *There exists a one-to-one and onto correspondence between submodules $\bar{S}$ of M/N and those S of M which contains N, defined by*

$$S = \pi^{-1}(\bar{S}) \qquad \text{and} \qquad \bar{S} = \pi(S).$$

Moreover, if S and $\bar{S}$ are corresponding submodules, then we have an isomorphism of R-modules:

$$V/S \xrightarrow{\sim} (M/N)/\pi(S).$$

PROOF. We leave the proof of this proposition as an exercise. □

1.3. Determinants. Matrices with entries in a commutative ring R can be manipulated in the same way as matrices with entries in a field F. This is because the operations of matrix addition and matrix multiplication are defined as in the case of matrices with entries in a field F; consequently they satisfy similar rules.

DEFINITION 3.8. *A matrix with entries in a commutative ring R is called an R-**matrix**.*

The determinant of an $n \times n$ R-matrix $A = (a_{ij})$ can be computed by any of the rules for matrices with entries in a field F. In particular, it is convenient to use the complete expansion,

$$\det(A) = \sum_p sgn(p) \cdot a_{1p(1)} \cdots a_{np(n)},$$

where the sum is taken over all permutations, i.e., bijections, of the set $\{1, 2, ..., n\}$, because it exhibits the determinant as a polynomial in the n^2 matrix entries a_{ij}.

The usual rules for determinant with entries in a field F apply, in particular

$$\det(AB) = \det(A)\det(B).$$

COROLLARY 3.3. *The invertible $n \times n$ matrices A with entries in a commutative ring R are those matrices whose determinants are units in R.*

PROOF. If A has a multiplicative inverse A^{-1} with entries in R, then

$$\det(A)\det(A^{-1}) = \det(I) = 1,$$

and hence the determinant of an invertible R-matrix is a unit, i.e., an invertible element, in R.

Conversely, if A is an R-module whose determinant $\det(A)$ is a unit, then we can compute its inverse using Cramer's Rule:

$$\det(A)I = A \cdot adj(A),$$

where the adjoint matrix $adj(A)$ is calculated from A by taking determinants of minors. This rule also holds in any ring, so

$$A^{-1} = adj(A)/\det(A).$$

$\square$

REMARK 3.2. *The determinant of an invertible matrix of size* 1×1 *with entries in* $\mathbb{Z}$ *is either* $+1$ *or* -1, *because there are only two invertible elements in* $\mathbb{Z}$. *But, if* $n > 1$, *then there are always reasonably many matrices, because elementary matrices of the type*

$$mulrow(i,u) = \begin{bmatrix} 1 & & & & & & \\ & \cdot & & & & & \\ & & 1 & & & & \\ & & & u & & & \\ & & & & 1 & & \\ & & & & & \cdot & \\ & & & & & & 1 \end{bmatrix},$$

where u *is a unit in* R, *are invertible, and elementary matrices of the type*

$$addrow(j,i,a) = \begin{bmatrix} 1 & & & & & & \\ & \cdot & & & & & \\ & & 1 & & & & \\ & & & \cdot & & & \\ & & & & \cdot & & \\ & & a & & & \cdot & \\ & & & & & & 1 \end{bmatrix} \tag{3.1}$$

$$\text{or} \quad \begin{bmatrix} 1 & & & & & & \\ & \cdot & & & a & & \\ & & \cdot & & & & \\ & & & \cdot & & & \\ & & & & 1 & & \\ & & & & & \cdot & \\ & & & & & & 1 \end{bmatrix} \tag{3.2}$$

where $i \neq j$ *and* $a \in R$ *have determinant* 1.

1.4. The matrices of homomorphisms of free R-modules. Let R^n be a free R-module and $\mathbf{B} = (v_1, ..., v_n)$ an ordered subset of an R-module M, and φ: $R^n \longrightarrow M$ be a map, defined by

$$X = \begin{bmatrix} x_1 \\ \cdot \\ \cdot \\ \cdot \\ x_n \end{bmatrix} \mapsto \mathbf{B}X = v_1x_1 + ... + v_nx_n.$$

Then one can easily check that

LEMMA 3.4. (1) *it is a homomorphism of R-modules, and*

(2) *this homomorphism is surjective if and only the set* $\mathbf{B} = (v_1, ..., v_n)$ *generates M, and*

(3) *the homomorphism is injective if and only if the set* $\mathbf{B} = (v_1, ..., v_n)$ *is (linearly) independent.* □

Thus, the homomorphism is an isomorphism if and only if $\mathbf{B} = (v_1, ..., v_n)$ is a basis of M, in which case M is a free module. Thus, a finitely generated R-module M has a basis if and only if it is free. As we have seen before, most finitely generated modules have no bases.

Computation with bases of free R-modules can be done in much the same way as with bases of vector spaces, using matrices with entries in R. In particular, we can speak of the **coordinate vector** $X = (x_1, ..., x_n)^t$ in R^n of an element $v \in M$, with respect to a basis $\mathbf{B} = (v_1, ..., v_n)$ of M. It is the unique column vector $X \in R^n$ such that

$$v = \mathbf{B}X = v_1x_1 + ... + v_nx_n.$$

Now, if two bases $\mathbf{B} = (v_1, ..., v_n)$ (old basis) and $\mathbf{B}' = (v_1', ..., v_r')$ (new basis) for the same free module M are given, then the **matrix of change of basis** is obtained as usual by writing the elements v_j of the old basis $\mathbf{B}$ as linear combination of the new basis:

$$\mathbf{B} = \mathbf{B}'P,$$

or,

$$v_j = \sum_{j=1}^{r} v_i' p_{ij}.$$

As with vector spaces, we have

PROPOSITION 3.2. *Any two bases of the same free finitely generated module M over a non-zero commutative ring R have the same cardinality.*

PROOF. Let $\mathbf{B}$ and $\mathbf{B}'$ be two bases of a free R-module M. We write elements of $\mathbf{B}$ as a linear combination of elements of $\mathbf{B}'$ and we write element of $\mathbf{B}'$ as a linear combination of elements of $\mathbf{B}$:

$$\mathbf{B} = \mathbf{B}'P \qquad \text{and} \qquad \mathbf{B}' = \mathbf{B}Q.$$

Thus, we have

$$\mathbf{B} = \mathbf{B}QP.$$

Then since $\mathbf{B}$ is a basis, there exists only one way to write each v_j as a linear combination of the v_j's:

$$v_j = 1v_j,$$

or $\mathbf{B} = \mathbf{B}I$; consequently we see that $QP = I$, and similarly $PQ = I$ holds.

We now note that P is an $r \times n$ matrix and Q is an $n \times r$ matrix, so we need to prove that $r = n$, in order to conclude that P and Q are indeed invertible R-matrices. For this, we suppose that $r > n$. Then we can always make P and Q square matrices by adding enough zeros:

$$I = PQ = \begin{bmatrix} P & 0 \end{bmatrix} \begin{bmatrix} Q \\ 0 \end{bmatrix}.$$

But then the determinant of the product PQ is equal to 1, whereas the product of the determinants, 0's, of these new square matrices with enough zeros is zero, which is impossible. Thus, we must always have $r = n$. □

DEFINITION 3.9. *The number of elements of a basis for a finitely generated free R-module M is call the **rank** of M, instead of the dimension.*

As in the case of vector spaces, we have

COROLLARY 3.4. *Every homomorphism* $\varphi\colon R^n \longrightarrow R^m$ *of free R-modules is left multiplication by an R-matrix.* $\square$

DEFINITION 3.10. *If* $\varphi\colon M \longrightarrow N$ *is a homomorphism of free R-modules with bases* $\mathbf{B} = (v_1, ..., v_n)$ *for* M *and* $\mathbf{C} = (w_1, ..., w_m)$ *for* N, *then the* ***matrix of the homomorphism*** φ *with respect to* $\mathbf{B}$ *and* $\mathbf{C}$ *is defined to be* $A = (a_{ij})$, *where*

$$\varphi(v_j) = \sum_{i=1}^{m} w_i a_{ij}, \qquad j = 1, 2, ..., n.$$

REMARK 3.3. *Thus, as in the case of linear maps for vector spaces, if* $\mathbf{B} = (v_1, ..., v_n)$ *is the old basis of* M, $\mathbf{B}' = (v_1', ..., v_n')$ *is a new basis of* M, *and* $\mathbf{C} = (w_1, ..., w_m)$ *is the old basis of* N, $\mathbf{C}' = (w_1', ..., w_m')$ *is a new basis of* N, *then the change of basis from the old basis to the new basis for* M *and* N, *respectively, gives us a new matrix* A' *of the homomorphism* φ *with respect to* $\mathbf{B}'$ *and* $\mathbf{C}'$:

$$A' = QAP^{-1},$$

where invertible R-matrices P *and* Q *are determined by the relations*

$$\mathbf{B} = \mathbf{B}'P \qquad \textit{and} \qquad \mathbf{C} = \mathbf{C}'Q.$$

Exercises for Section 3.1

EXERCISE 3.1. *Verify that any commutative ring* R *is a module over itself.*

EXERCISE 3.2. *Let* R *be a ring, considered as an R-module. How many module homomorphisms* $\varphi\colon R \longrightarrow R$ *we can have?*

EXERCISE 3.3. *Let* $\varphi\colon V \longrightarrow W$ *be a homomorphism of modules over a (commutative) ring* R, *and let* V' *and* W' *be submodules of* V *and* W, *respectively. Prove that* $\varphi(V')$ *is a submodule of* W *and* $\varphi^{-1}(W')$ *is a submodule of* V.

EXERCISE 3.4. *An R-module is called* ***simple*** *if it has no non-trivial proper submodule.*

(1) *Prove that any simple module if isomorphic to R/M, where M is a maximal ideal of R.*
(2) *Prove Schur's Lemma: Let $\varphi\colon S \longrightarrow S'$ be a homomorphism of simple modules. Then either φ is the zero map or else it is an isomorphism.*

EXERCISE 3.5. *Let R be a ring, V an R-module, and let $End(V)$ the set of all endomorphisms of V. Prove that $End(V)$ is a non-commutative ring under composition of module homomorphisms as multiplication and addition defined by $(\varphi + \phi)(m) = \varphi(m) + \phi(m)$.*

EXERCISE 3.6. *Prove that $End(V)$ of simple module V is a field.*

EXERCISE 3.7. *The annihilator of an R-module V is by definition the subset I of R:*

$$I = \{r \in R : rv = 0 \textit{ for each } v \in V\}.$$

(1) *Prove that I is an ideal of R.*
(2) *What is the annihilator of the $\mathbb{Z}$-module $\mathbb{Z}/2\mathbb{Z} \times \mathbb{Z}/3\mathbb{Z} \times \mathbb{Z}/4\mathbb{Z}$?*

EXERCISE 3.8. *Let I be an ideal of a ring R. Prove that I is a free R-module if and only if it is a principal ideal, generated by an element α which is not a zero-divisor in R.*

EXERCISE 3.9. *Prove that every subgroup of $\mathbb{Z}$ is a free $\mathbb{Z}$-module.*

EXERCISE 3.10. *Let A be the matrix of a homomorphism $\varphi\colon \mathbb{Z}^n \longrightarrow \mathbb{Z}^m$ between free modules.*

(1) *Prove that φ is injective if and only if the rank of A is n.*
(2) *Prove that φ is surjective if and only if the greatest common divisor of the determinants of the $m \times m$ minors of A is 1.*

2. Diagonalization of Integer Matrices

The main purpose of this section is about how to simplify an $m \times n$ integer matrix $A = (a_{ij})$ by a succession of elementary operations, using possibly elementary row operations and column operations. The same method will work for matrices with entries in a Euclidean domain (ED) and, with some modification, for matrices with entries in a principal ideal domain (PID). We need this procedure to classify abelian groups in a later section of this chapter.

DEFINITION 3.11. *An integral domain R is called a* ***Euclidean domain*** *if there exists a size function $\sigma\colon R - \{0\} \longrightarrow \mathbb{N} \cup \{0\}$ such that the division algorithm holds:*

*(**Division Algorithm**) Let $a \neq 0, b \in R$. Then there exist elements $q, r \in R$ such that $b = aq + r$ with either $r = 0$ or $\sigma(r) < \sigma(a)$.*

REMARK 3.4. *We do not require the elements q, r to be uniquely determined by a and b.*

EXAMPLE 3.5.
(1) $a \longmapsto |a|$ *(the absolute value of a) is the size function: $\mathbb{Z} - \{0\} \longrightarrow \mathbb{N} \cup \{0\}$.*
(2) $f(x) \longmapsto \deg(f(x))$ *(the degree of $f(x)$) is the size function: $F[x] - \{0\} \longrightarrow \mathbb{N} \cup \{0\}$.*
(3) $a \longmapsto |a|^2$ *(the square of the absolute value of a) is the size function: $\mathbb{Z}[i] - \{0\} \longrightarrow \mathbb{N} \cup \{0\}$, where $i = \sqrt{-1}$.*

We are going to apply to an integer matrix the following elementary operations:

(1) Add an integer multiple of one row to another, or add an integer multiple of one column to another;
(2) interchange two rows or two columns;
(3) multiply a row or a column by a unit u.

REMARK 3.5. *In contrast to those elementary row operations and column operations for matrices with entries in a field F, the scalar u in the* *third operation must be a unit, i.e., an invertible element of R, otherwise, the third operation is not going to have a corresponding inverse operation.*

THEOREM 3.3. *Let A be an $m \times n$ integer matrix. Then there exist products Q, P of elementary integer matrices as above, so that $A' = QAP^{-1}$ is a diagonal matrix:*

$$D = \begin{bmatrix} \begin{bmatrix} d_1 & & \\ & \ddots & \\ & & d_r \end{bmatrix} & \\ & 0 \end{bmatrix}, \tag{3.3}$$

where the diagonal entries d_i are non-negative and $d_1|d_2, d_2|d_3, \ldots$, that is, each diagonal entry d_i divides d_{i+1}.

REMARK 3.6. *As the following examples will show, Q and P in the theorem are not uniquely determined.*

PROOF. Our strategy is to perform a sequence of operations until we end up with a matrix:

$$\begin{bmatrix} d_1 & 0 \\ 0 & \begin{bmatrix} & B & \end{bmatrix} \end{bmatrix} \tag{3.4}$$

where d_1 divides each entry of B. When this is done, we work on B, inductively. The process is based on repeated division with remainder. We may assume that $A \neq 0$.

(Step 1) By permuting rows and columns, we move a non-zero entry with smallest absolute value to the upper left corner. Multiply the first row by -1 if necessary, so that this upper left entry a_{11} becomes positive.

We now try to clear out the first row and column. Whenever an operation produces a non-zero entry in the matrix whose absolute value is smaller than a_{11}, we go back to (Step 1) above and start the whole process over, and this process guarantees to reduce the size of a_{11} each time; hence we will not have to return (Step 1) infinitely often.

(Step 2) We choose a non-zero entry a_{i1} in the first column, with $i > 1$, and divide by a_{11}:

$$a_{i1} = a_{11}q + r,$$

where $0 \leq r < a_{11}$. Thus, we subtract q times (row 1) from (row i) in order to change a_{i1} with r.

If $r \neq 0$, then we go bake to (Step 1).

Otherwise, i.e., if $r = 0$, then we have produced a zero in the first column.

Finitely many repetitions of (Step 1) and (Step 2) result in a matrix in which $a_{i1} = 0$ for each $i > 1$.

Similarly, we may use the analogue of (Step 2) for column operations to clear out the first row, eventually ending up with a matrix in which the only non-zero entry in the first row and column is a_{11}, as required. However, a_{11} may not yet divide each entry of the matrix B.

(Step 3) Assume that a_{11} is the only non-zero entry in the first row and column, but that some entry b of B is not divisible by a_{11}. Then add the column of A which contains b to the first column to have an entry b in the first column.

We then go back to (Step 2). Division with remainder will now produce a smaller matrix entry, sending us back to (Step 1).

A finite sequence of these steps will produce a matrix of the form (3.4, which then allows us to proceed by induction. □

EXAMPLE 3.6. *Let*

$$A = \begin{bmatrix} 2 & -1 \\ 1 & 2 \end{bmatrix}.$$

We try to reduce A in two different ways:

$$\begin{bmatrix} 2 & -1 \\ 1 & 2 \end{bmatrix} \overset{\text{row}}{\longrightarrow} \begin{bmatrix} 1 & 2 \\ 2 & -1 \end{bmatrix} \overset{\text{row}}{\longrightarrow} \begin{bmatrix} 1 & 2 \\ 0 & -5 \end{bmatrix} \overset{\text{col}}{\longrightarrow} \begin{bmatrix} 1 & 0 \\ 0 & -5 \end{bmatrix} \overset{\text{row}}{\longrightarrow} \begin{bmatrix} 1 & 0 \\ 0 & 5 \end{bmatrix},$$

and hence $P^{-1} = \begin{bmatrix} 1 & -2 \\ 0 & 1 \end{bmatrix}$ *and* $Q = \begin{bmatrix} 1 & 0 \\ 0 & -1 \end{bmatrix}\begin{bmatrix} 1 & 0 \\ -2 & 1 \end{bmatrix}\begin{bmatrix} 0 & 1 \\ 1 & 0 \end{bmatrix}$, *so that*

$$A' = QAP^{-1} = \begin{bmatrix} 1 & 0 \\ 0 & 5 \end{bmatrix}.$$

Now secondly,

$$\begin{bmatrix} 2 & -1 \\ 1 & 2 \end{bmatrix} \overset{\text{col}}{\to} \begin{bmatrix} 1 & -1 \\ 3 & 2 \end{bmatrix} \overset{\text{col}}{\to} \begin{bmatrix} 1 & 0 \\ 3 & 5 \end{bmatrix} \overset{\text{row}}{\to} \begin{bmatrix} 1 & 0 \\ 0 & 5 \end{bmatrix},$$

and hence $Q = \begin{bmatrix} 1 & 0 \\ -3 & 1 \end{bmatrix}$, $P^{-1} = \begin{bmatrix} 1 & 0 \\ 1 & 1 \end{bmatrix}\begin{bmatrix} 1 & 1 \\ 0 & 1 \end{bmatrix}$, *so that*

$$A' = QAP^{-1} = \begin{bmatrix} 1 & 0 \\ 0 & 5 \end{bmatrix}.$$

Thus, we do not need to follow the above systematic method to obtain A'.

We note that the key ingredient in the above proof is the *division algorithm*, and hence the same proof will work when $\mathbb{Z}$ is replaced with any Euclidean domain.

THEOREM 3.4. *Let* R *be a Euclidean domain, for example a polynomial ring* $F[t]$ *in one variable over a field* F. *Let* A *be an* $m \times n$ *matrix with entries in* R. *Then there exist products* Q, P *of elementary* R*-matrices such that* $A' = QAP^{-1}$ *is a diagonal matrix and each diagonal entry of* A' *divides the next, i.e.,* $d_1|d_2|d_3|\ldots$, *and each* d_j *is non-negative. If* $R = F[t]$, *then we can moreover normalize by requiring the polynomials* d_i *to be monic.* □

EXAMPLE 3.7. *Here is a diagonalization of a matrix with entries consisting of integer polynomials. Let*

$$A = \begin{bmatrix} (t-2)(t-1) & t-2 \\ (t-1)^3 & (t-2)(t-1) \end{bmatrix}.$$

This example will tell us something. If one attempts to reduce entries to zeros too quickly like:

$$A \overset{\text{col}}{\to} \begin{bmatrix} 0 & t-2 \\ (t-1)^2 & (t-2)(t-1) \end{bmatrix} \overset{\text{row}}{\to} \begin{bmatrix} 0 & t-2 \\ (t-1)^2 & 0 \end{bmatrix}$$

$$\overset{\text{col}}{\to} \begin{bmatrix} t-2 & 0 \\ 0 & (t-1)^2 \end{bmatrix},$$

but $t-2$ never divides $(t-1)^2$; hence the final matrix is not what we are looking for. The point we missed in this computation is that we divided out other entries with $t-2$ without first checking that $t-2$ is the smallest one we should have during our reduction process on the entries of A.

In the following computation, we will see that $1 = \gcd(t-1, t-2)$, the greatest common divisor *of $t-1$ and $t-2$, is in fact the smallest one. Since $(t-1)^3 = t^3 - 3t^2 + 3t - 1 = t \cdot (t-2)(t-1) + (t-1)$ by the division algorithm,*

$$A \overset{\text{row}}{\to} \begin{bmatrix} (t-2)(t-1) & t-2 \\ t-1 & -(t-2) \end{bmatrix} \overset{\text{row}}{\to} \begin{bmatrix} (t-1) & -(t-2) \\ (t-2)(t-1) & t-2 \end{bmatrix}$$

$$\overset{\text{col}}{\to} \begin{bmatrix} 1 & -(t-2) \\ (t-2)t & t-2 \end{bmatrix} \overset{\text{row}}{\to} \begin{bmatrix} 1 & -(t-2) \\ 0 & (t-2)(t-1)^2 \end{bmatrix}$$

$$\overset{\text{col}}{\to} \begin{bmatrix} 1 & 0 \\ 0 & (t-2)(t-1)^2 \end{bmatrix},$$

Thus, $A' = \begin{bmatrix} 1 & 0 \\ 0 & (t-2)(t-1)^2 \end{bmatrix}.$

The diagonalization of integer matrices can be used to describe homomorphisms between free abelian groups, i.e., free $\mathbb{Z}$-modules.

THEOREM 3.5. *Let $\varphi\colon M \longrightarrow N$ be a homomorphism of free abelian groups. Then there exist bases of M and N such that the matrix of the homomorphism φ with respect to these bases is of the diagonal form in Theorem 3.3.*

PROOF. As we have already described in Remark 3.3, a homomorphism φ: $M \longrightarrow N$ of free abelian groups is described by a matrix A once bases for M and N are chosen. Then by Theorem 3.3 there exists a change of basis by invertible integer matrices P, Q which changes A to a diagonal matrix $A' = QAP^{-1}$. □

Let φ: $M \longrightarrow N$ be a homomorphism of R-modules. Then we know that the kernel of φ, $\ker(\varphi)$, and the image of φ, $img(\varphi)$, are submodules of M and N, respectively.

REMARK 3.7. *Our general concern is that under what conditions they are finitely generated if both M and N are finitely generated R-modules. In particular, in the next section we are going to prove that every subgroup, i.e., $\mathbb{Z}$-submodule, of a finitely generated abelian group, i.e., $\mathbb{Z}$-module, is finitely generated. (This is because the ring $\mathbb{Z}$ is a Noetherian ring, as we shall see in the next section. See Proposition 3.5.) Before we move into a more general setting, we can say something when both M and N are free R-modules of finite rank, respectively.*

Let φ: $\mathbb{Z}^n \longrightarrow \mathbb{Z}^m$ be left multiplication by an $m \times n$ integer matrix A. The kernel of φ is the subgroup of $\mathbb{Z}^n$ which consists of the system of linear equations:

$$AX = 0. \tag{3.5}$$

Then by the above theorem, we can diagonalize A, say to $A' = QAP^{-1}$, where Q, P are products of elementary integer matrices. To solve (3.5), we make the change of variable $X' = PX$ and solve the diagonal system

$$A'X' = QAP^{-1}X' = 0,$$

say, $d_1x'_1 = 0, ..., d_nx'_n = 0$. If $d_i = 0$ then x'_i is arbitrary, and if $d_i \neq 0$, then x'_i must be zero. Here $X = (x_1, ..., x_n)^t$ and $X' = (x'_1, ..., x'_n)^t$, as usual. Thus, we have solved $QAX = 0$. Then since Q is invertible, both $QAX = 0$ and $AX = 0$ have the same solutions, $X = P^{-1}X'$ of the original system.

Now, what is the image of left multiplication by an $m \times n$ integer matrix $A = \varphi$: $\mathbb{Z}^n \longrightarrow \mathbb{Z}^m$? By definition, it consists of vectors $B \in \mathbb{Z}^m$ such that the system of linear equations $AX = B$ has an integer solution. Since the image of each element e_i of the standard basis $(e_1, ..., e_n)$ of the domain $\mathbb{Z}^n$ is mapped under A to

$$Ae_i = A_i$$

where $A = [A_1, ..., A_n]$, each $B = AX = A_1x_1 + ... + A_nx_n$ for some x_i's. That is,

COROLLARY 3.5. *The image of left multiplication by A is generated by the columns of A as a $\mathbb{Z}$-submodule of $\mathbb{Z}^m$.*

Having said these as warm-up, we now, conversely, assume that we are given a finite generating set $A_1, ..., A_n \in \mathbb{Z}^m$ of a subgroup S of a free abelian group $\mathbb{Z}^m$. Then each generator A_i is the image of the free generator, i.e., the basis element, e_i of $\mathbb{Z}^n$ under left multiplication by the matrix $A := [A_1, ..., A_n]$; hence S is just the image of this left multiplication by A: $\mathbb{Z}^n \longrightarrow \mathbb{Z}^m$.

DEFINITION 3.12. *Such a matrix A is said to* ***represent*** *the generating set $(A_1, ..., A_n)$.*

In fact, this construction is universal by Theorem 3.1, and this interpretation of S as an image of a homomorphism can tell the meaning of left and right multiplication to A by invertible integer matrices Q and P^{-1}:

(1) Left multiplication $QA = Q[A_1, ..., A_n]$ to A by Q corresponds to a change of basis in the module $\mathbb{Z}^m$, the range of the map, and its effect is to multiply each of the generators A_i by the same quantity Q: $QA = [QA_1, ..., QA_n]$.
(2) Right multiplication AP^{-1} to A by P^{-1} represents a change of basis in the domain $\mathbb{Z}^n$, and its effect is to change the generating set $[A_1, ..., A_n]$ of S by a column operation P^{-1}: $AP^{-1} = [A_1, ..., A_n]P^{-1}$.

EXAMPLE 3.8. *For example,* $[A_1, A_2]\begin{bmatrix} 1 & r \\ 0 & 1 \end{bmatrix} = [A_1, A_2 + rA_1]$.

Combining these observations with diagonalization results in the following theorem:

THEOREM 3.6. *Let S be a subgroup of a free abelian group M of rank m. Then there exists a basis $(w_1, ..., w_m)$ of M and a basis $(u_1, ..., u_n)$ of S with the following properties:*

(1) $n \leq m$,
(2) *for each $j \leq n$ there exists a positive integer d_j such that $u_j = d_j w_j$, and*
(3) $d_1 | d_2 | d_3 | \ldots$.

PROOF. By Proposition 3.5 which we are going to prove later, S is finitely generated as a $\mathbb{Z}$-submodule of M. Thus, let $(w_1, ..., w_m)$ be a basis $\mathbf{B}$ of M and $(u_1, ..., u_n)$ be a set of generators for S, which gives us an $m \times n$ matrix A which represents S as above. Then by the diagonalization theorem (that is, either one of Theorems 3.3, 3.4, or 3.5, there exist invertible integral matrices P, Q such that

$$A' = QAP^{-1}$$

is a diagonal matrix with diagonal entries d_j's with $d_j | d_{j+1}$ for each j. This exactly means that the old basis of the domain is changed to a new basis by P and there exists a new basis $\mathbf{B}' = (w_1', ..., w_m')$ of M such that $\mathbf{B} = \mathbf{B}'Q$ and the old generating set $(u_1, ..., u_n)$ of S is changed to a new generating set $(u_1', ..., u_n')$ of S. Then since the diagonal matrix

$$A' = \begin{bmatrix} \begin{bmatrix} d_1 & & \\ & \ddots & \\ & & d_n \end{bmatrix} & 0 \\ 0 & 0 \end{bmatrix},$$

where $d_j|d_{j+1}$ for each $j = 1, ..., n$ and each d_j is non-negative, represents the new generating set $(u'_1, ..., u'_n)$ of S, we have that each

$$u'_j = d_j w'_j.$$

This is because, since u'_j is the image of the jth basis element e'_j of the new basis of the domain by Theorem 3.1 for instance, which is exactly the same as the unique linear combination of the basis elements w'_i's with coefficients in the jth column of the diagonal matrix A':

$$[u'_1, ..., u'_n] = [A'(e'_1), ..., A'(e'_n)] = (w'_1, ..., w'_m)A'.$$

We then drop the primes to obtain the basis and generating set required, and the third condition $d_1|d_2|d_3|...$ is established.

Now, if $n > m$ then since A' is diagonal as above, its jth columns are all zero for $j > m$; hence each $u_j = 0w_1 + ... + 0w_m = 0$ for $j > m$. Since the zero element is useless as a generator we just throw it out of the generating set $(u_1, ..., u_n)$. For the same reason, we may throw out a generator u_j whenever $d_j = 0$; consequently, all the d_j will be positive and we will have $n \leq m$.

Finally, we need to verify that if the basis and generating sets are chosen so that $d_j > 0$ and $n \leq m$, then $(u_1, ..., u_n)$ is a basis of S. But, this very easy: Since it generates S, we only need to prove that $(u_1, ..., u_n)$ is (linearly) independent. Assume that $r_1u_1 + ... + r_nu_n = 0$. Then since each $u_j = d_jw_j$ and since $(w_1, ..., w_m)$ is a basis, each $r_jd_j = 0$; hence each $r_j = 0$ since each $d_j > 0$. This finishes the proof of the whole theorem. □

REMARK 3.8. *We note that if S is the zero subgroup, then we will end up in the above proof throwing out all the generators. To avoid this, we must adopt the convention that the empty set generates the zero module, as with vector spaces. Or else, we need to make a special mention of this exceptional case in the statement of the theorem.*

COROLLARY 3.6. *Every subgroup of a free abelian group of rank m is free, and its rank is at most m.* □

Theorem 3.6 is quite explicit: Let S be the subgroup of $\mathbb{Z}^m$ generated by the columns of a matrix A, and suppose that $A' = QAP^{-1}$ is a diagonal matrix in either one of Theorems 3.3, 3.4, or 3.5. To display S in the form asserted in Theorem 3.6, we rewrite this equation in the form

$$Q^{-1}A' = AP^{-1}. \tag{3.6}$$

Then as we have already seen right before Theorem 3.6, the columns of the matrix AP^{-1} form our new set of generators for S. Then since

$$A' = \begin{bmatrix} \begin{bmatrix} d_1 & & \\ & \ddots & \\ & & d_n \end{bmatrix} & 0 \\ 0 & 0 \end{bmatrix}$$

is diagonal, (3.6) also tells us that the new generators are multiples of the columns of Q^{-1}. We change the basis of $\mathbb{Z}^m$ from the standard basis $(e_1, ..., e_m)$ to the basis consisting of the columns of Q^{-1}, because the matrix of this change of basis is Q. Then the new generators u'_j are d_j multiples of the new basis elements w'_j.

EXAMPLE 3.9. *Let S be the subgroup of $\mathbb{Z}^2$ generated by the two columns of the matrix of Example 3.6. By the computation we have done there, we have:*

$$\begin{aligned} Q^{-1}A' &= \begin{bmatrix} 1 & 0 \\ 3 & 1 \end{bmatrix}\begin{bmatrix} 1 & 0 \\ 0 & 5 \end{bmatrix} = \begin{bmatrix} 1 & 0 \\ 3 & 5 \end{bmatrix}, \\ AP^{-1} &= \begin{bmatrix} 2 & -1 \\ 1 & 2 \end{bmatrix}\begin{bmatrix} 1 & 1 \\ 1 & 2 \end{bmatrix} = \begin{bmatrix} 1 & 0 \\ 3 & 5 \end{bmatrix}. \end{aligned}$$

Thus, we have $[w'_1, w'_2]A' = [u_1, u_2]P^{-1} = [u'_1, u'_2]$. This means that the new basis $(w'_1, w'_2) = \left(\begin{bmatrix} 1 \\ 3 \end{bmatrix}, \begin{bmatrix} 0 \\ 1 \end{bmatrix}\right)$, and the new generators of S are

$$[u'_1, u'_2] = [u_1, u_2]P^{-1} = [w'_1, w'_2]A' = [w'_1, w'_2]\begin{bmatrix} 1 & 0 \\ 0 & 5 \end{bmatrix} = (w'_1, 5w'_2).$$

Exercises for Section 3.2

EXERCISE 3.11. *Find a basis for the following submodules of* $\mathbb{Z}^3$*:*

(1) *The module generated by* $(1,0,-1),(2,-3,1),(0,3,1),(3,1,5)$.

(2) *The module of solutions of the system of equations*

$$\begin{cases} x+2y+3z &= 0 \\ x+4y+9z &= 0 \end{cases}$$

EXERCISE 3.12. *Reduce each of the following integer matrices to diagonal form by integer row and column operations:*

(1) $\begin{bmatrix} 3 & 1 \\ -1 & 2 \end{bmatrix}$.

(2) $\begin{bmatrix} 1 & 2 & 3 \\ 4 & 5 & 6 \end{bmatrix}$.

(3) $\begin{bmatrix} 3 & 1 & -4 \\ 2 & -3 & 1 \\ -4 & 6 & -2 \end{bmatrix}$.

EXERCISE 3.13. *Let* $A = \begin{bmatrix} 4 & 7 & 2 \\ 2 & 4 & 6 \end{bmatrix}$.

(1) *Determine integer matrices* P^{-1} *and* Q *such that* QAP^{-1} *is a diagonal matrix.*

(2) *Determine all integer solutions to the system of equations* $AX = 0$.

3. Generators and Linear Relations

3.1. Presentation matrices. In the previous section, we have studied homomorphisms between free $\mathbb{Z}$-modules of finite rank and we have proved that every $\mathbb{Z}$-submodule of a free $\mathbb{Z}$-module M of rank m is again free of finite rank $\leq m$. (See Corollary 3.6). In this section, we turn our attention to modules which are *not* free. Then, in the next section we are going to prove Proposition 3.5 which was needed in the proof of Theorem 3.6.

REMARK 3.9. *The main concern of this section can be heuristically explained as follows: One of the main purposes of Linear Algebra is to learn how to solve a system of linear equations where two different quantities appear, the known ones and the unknown ones, and then we study related subjects, showing up in many different parts of Mathematics, pure and/or applied, and other parts of Sciences near Mathematics.*

Here, what we should note is that those quantities called known ones can be actually unknown to us and all really matters are the fact that we consider certain symbols to be known to us and other symbols not yet to be known to us. That is, we regard certain quantities to be known to us and other quantities not yet known to us, in order for us to easily approach or handle them when we try or attempt to mathematically understand the subject under consideration.

This clearly means that their role as "known" things or "unknown" things can be interchanged, and this approach is already utilized in the theory of invariants of the late 19th *century. On the other hand, for Hardware called Computers and Software called Computer Algebra Systems (CAS), considering either of them as known or unknown is quite arbitrary.*

Thus, we would like to study what's going on among the known ones, i.e., the coefficients or the column vectors consisting of them, when the symbols for the unknown ones are anything. That is, when a system of linear relations among generators is given to us, we forget about the generators themselves by regarding them simply as symbols, but we would like to study how the column vectors of coefficients are linearly related, and if so, we would like to figure out in detail the "smallest or linearly independent" generating set of column vectors which generates the whole column vectors of coefficients.

In the next section, we will then apply those results developed in this section and the diagonalization procedure of the previous section to matrices in order to study the structure of abelian groups.

We start this section with an example:

EXAMPLE 3.10. *We consider an abelian group or* $\mathbb{Z}$*-module* M *generated by three elements* (v_1, v_2, v_3)*, satisfying the following relations:*

$$\begin{aligned} 3v_1 + 2v_2 + v_3 &= 0 \\ 8v_1 + 4v_2 + 2v_3 &= 0 \\ 7v_1 + 6v_2 + 2v_3 &= 0 \\ 9v_1 + 6v_2 + v_3 &= 0 \end{aligned} \tag{3.7}$$

We can rewrite this into matrix multiplication,

$$(v_1, v_2, v_3)A = (0, 0, 0, 0),$$

where

$$A = \begin{bmatrix} 3 & 8 & 7 & 9 \\ 2 & 4 & 6 & 6 \\ 1 & 2 & 2 & 1 \end{bmatrix} \tag{3.8}$$

is an 3×4 *integer matrix whose columns consist of the integer coefficients of the relations (3.7), and hence it will define a linear homomorphism:*

$$\mathbb{Z}^4 \longrightarrow \mathbb{Z}^3,$$

as is explained in the previous section, right after Theorem 3.5. We now need a few more terminologies:

DEFINITION 3.13. (1) *Equations of the form*

$$a_1 v_1 + \ldots + a_m v_m = 0, \qquad a_i \in R \tag{3.9}$$

are called ***relations*** *among the elements* $(v_1, \ldots, v_m)$.

(2) *The* R*-vector* $(a_1, \ldots, a_m)^t$ *which determine the relation is called a* ***relation vector****.*

(3) *A* ***complete set of relations*** *is a set of relation vectors such that every relation vector is a linear combination of relation vectors in this set.*

Thus, our concern is how to figure out a complete set of relations from the given matrix A, because a matrix such as (3.8) will not describe the module M completely unless its columns form a complete set of relations.

As we shall see it soon, it becomes much easier for us to understand the concept of a complete set of relations if we work with homomorphism of free modules rather than directly with the relations or the relation vectors: Let A be an $m \times n$ matrix with entries in a commutative ring R. As we have already seen it, left multiplication by A is a homomorphism of R-modules

$$\varphi\colon R^n \longrightarrow R^m. \tag{3.10}$$

We know that the kernel of φ and the image of φ are the two natural objects we usually consider whenever we are given a homomorphism

$$\varphi : M \longrightarrow M'$$

of R-modules, but there is one more natural object we need to consider, which is usually called the **cokernel** of φ and defined by the quotient module:

$$M'/(im\ \varphi). \tag{3.11}$$

REMARK 3.10. *In categorical terminologies, the kernel of φ is the "subobject" of φ and the cokernel of φ is its dual, "quotient object" of φ. And we always have these two objects in the category of R-modules.*

If we denote the image of left multiplication by A by AR^n as usual, the cokernel of (3.10) is

$$R^m/AR^n,$$

and this cokernel is usually said to **be represented by the matrix** A. More generally, we have

DEFINITION 3.14. *Any isomorphism*

$$\sigma\colon R^m/AR^n \xrightarrow{\sim} M \tag{3.12}$$

is called a ***representation*** *of an R-module M, and we say that the matrix A is a* ***presentation matrix*** *for M if there is such an isomorphism.*

EXAMPLE 3.11. *The cyclic group $\mathbb{Z}/5\mathbb{Z}$ is presented as a $\mathbb{Z}$-module by the 1×1 integer matrix $[5]$. On the other hand, consider the $\mathbb{Z}$-module*

M presented by the matrix

$$\begin{bmatrix} 2 & -1 \\ 1 & 2 \end{bmatrix}$$

is generated by two elements v_1 and v_2 with relations vectors $(2,1)^t$ and $(-1,2)^t$:

$$\begin{array}{rcl} 2v_1 + v_2 & = & 0 \\ -v_1 + 2v_2 & = & 0 \end{array}.$$

It seems that the module requires two generators, but it is not the case. In this simple case, we can easily solve the first equation to see that $v_2 = -2v_1$, which is a linear combination of v_1, and substituting this into the second equation we obtain $-5v_1 = 0$. This means that the module M is generated by a single generator v_1, with a single relation $5v_1 = 0$. Thus, M is isomorphic to $\mathbb{Z}/5\mathbb{Z}$; hence this 2×2 matrix also represents the cyclic group $\mathbb{Z}/5\mathbb{Z}$.

Here is a systematic and theoretical method of finding a presentation of a given R-module M: Let M be a finitely generated R-module and so let $(v_1, ..., v_m)$ be its generating set, an ordered set. Then by Theorem 3.1, we have a canonical surjective homomorphism of R-modules

$$f\colon R^m \longrightarrow M,$$

which sends the column vector $(x_1, ..., x_m)^t = \sum_{i=1}^m x_i e_i \mapsto \sum_{i+1}^m x_i v_i = v_1 x_1 + ... + v_m x_m$, that is, sending each of the generating elements, the standard basis elements, e_i of R^m to each of the generating elements v_i. If we denote its kernel by W, we have by the first isomorphism theorem in Proposition 3.1 that M is isomorphic to R^m/W.

If the submodule W of R^m is not finitely generated, then we cannot go further anymore.

But if W is again a finitely generated R-module like the module M, say by a generating set $(w_1, ..., w_n)$, an order set, then we can repeat the above procedure and we have another surjective homomorphism

$$g\colon R^n \longrightarrow W,$$

as before, and hence the composition of g with the obvious inclusion homomorphism i: $W \hookrightarrow R^m$ gives us a homomorphism of R-modules,

$$\varphi\colon R^n \longrightarrow R^m,$$

which sends each element e_i of the standard basis $(e_1, ..., e_n)$ of R^n to w_i, an element of the generating set $(w_1, ..., w_n)$ of W.

$$e_i \longmapsto w_i = \begin{bmatrix} a_{1i} \\ \vdots \\ a_{mi} \end{bmatrix}, \qquad i = 1, ..., n$$

Thus, φ is left multiplication by the $m \times n$ matrix A consisting of w_i's. This is because, for any column vector $(x_1, ..., x_n)^t$ of R^n, we have

$$\begin{aligned} \begin{bmatrix} x_1 \\ \vdots \\ x_n \end{bmatrix} = \textstyle\sum_{i=1}^n e_i x_i \longmapsto \sum_{i=1}^n w_i x_i \quad &= w_1 x_i + ... + w_n x_n \\ &= (w_1, ..., w_n) \begin{bmatrix} x_1 \\ \vdots \\ x_n \end{bmatrix} \\ &= A \begin{bmatrix} x_1 \\ \vdots \\ x_n \end{bmatrix}. \end{aligned}$$

By our construction, W is clearly the image of φ: $R^n \longrightarrow R^m$, and hence it is AR^n. Thus,

$$R^m/AR^n = R^m/W \xrightarrow{\sim} M,$$

that is, the $m \times n$ matrix A is a presentation matrix for M, where m is the number of generators of the module M and n the number of generators of the kernel of the surjective module homomorphism $R^m \longrightarrow M$. Since the w_i's generate W, these columns $w_i = (a_{1i}, ..., a_{mi})^t$, $i = 1, ..., n$ form a complete set of relations among the generators $(v_1, ..., v_m)$ of the module M giving us

$$(v_1, ..., v_n)A = 0.$$

Thus a presentation matrix A for a finitely generated R-module M is determined by

(1) A set of generators for M,
(2) a complete set of relations among these generators,

provided that the kernel W in the above explanation is a finitely generated R-module.

REMARK 3.11. *As we shall see it later, Proposition 3.5 will provide us a sufficient condition for it, namely, R being a Noetherian ring.*

In particular, we note that $\mathbb{Z}$ is a Noetherian ring because it is a PID (Principal Ideal Domain).

Since any presentation matrix depends on the choices (1) and (2) above, other matrices can present the same module, or isomorphic modules. Here are some rules for manipulating a matrix A without changing the isomorphism class of the module it presents. *The striking point is that the size of presentation matrix can be reduced*, which we have already seen in Example 3.11.

PROPOSITION 3.3. *Let A be an $m \times n$ presentation matrix for an R-module M. The following matrices A' present the same module M:*

(1) *$A' = QAP^{-1}$ for some invertible R-matrix Q of size $m \times m$ and some invertible R-matrix P of size $n \times n$;*
(2) *A' is obtained by deleting a column of zeros from A;*
(3) *A' is obtained from A by deleting the ith row and j^{th} column when the jth column of A is e_i.*

PROOF. (1) The module R^m/AR^n is isomorphic to M. Then since the change of A to QAP^{-1} corresponds to a change of basis in R^m and R^n, the isomorphism class of the quotient module does not change.

(2) A column consisting of zero corresponds to the trivial relation, $0v_1 + ... + 0v_n = 0$, which can be omitted.

(3) Suppose that the jth column of the matrix A is e_i. Then the corresponding relation is $1v_i = 0$ in M, and hence v_i can be left out of

the generating set $(v_1, ..., v_n)$ of M. Doing this changes the matrix A by deleting the ith row and j^{th} column of A. $\square$

Thus as the following example shows, we can simplify a matrix quite a lot by these rules.

EXAMPLE 3.12. *Let A be the integer matrix in our first example of this section, Example 3.10. We apply elementary row and column operations (without row exchanges) by (1) of the above proposition, and then we apply (2) or (3) of the above proposition to obtain A'.*

$$A = \begin{bmatrix} 3 & 8 & 7 & 9 \\ 2 & 4 & 6 & 6 \\ 1 & 2 & 2 & 1 \end{bmatrix} \to \begin{bmatrix} 0 & 2 & 1 & 6 \\ 0 & 0 & 2 & 4 \\ 1 & 2 & 2 & 1 \end{bmatrix} \mapsto \begin{bmatrix} 2 & 1 & 6 \\ 0 & 2 & 4 \end{bmatrix}$$

$$\mapsto \begin{bmatrix} 2 & 1 & 6 \\ -4 & 0 & -8 \end{bmatrix} \mapsto \begin{bmatrix} -4 & -8 \end{bmatrix} \mapsto [4].$$

Thus, A presents the abelian group $\mathbb{Z}/4\mathbb{Z}$.

By definition, an $m \times n$ matrix presents a module by means of m generators and n relations among m generators. But as we have seen from this example, the number of generators and the number of relations depend on choices we have made, and hence they are not uniquely determined by the given module M.

EXAMPLE 3.13. *The 2×1 matrix $\begin{bmatrix} 4 \\ 0 \end{bmatrix}$ presents a module an abelian group M by means of two generators (v_1, v_2) and one relation $4v_1 = 0$. That is, $(v_1, v_2)\begin{bmatrix} 4 \\ 0 \end{bmatrix} = 0$. Thus, v_2 can be any element and $4v_1 = 0$. Thus, the group which it presents is isomorphic to the product group $\mathbb{Z}/4\mathbb{Z} \times \mathbb{Z}$. Here, since $4 \neq 1$, it is not an e_1.*

On the other hand, the 1×2 matrix $\begin{bmatrix} 4 & 0 \end{bmatrix}$ presents a group with one generator v_1 and two relations, of which the second relation is trivial relation $0v_1 = 0$, thus $\begin{bmatrix} 4 & 0 \end{bmatrix}$ is reduced to $[4]$; hence it is isomorphic to $\mathbb{Z}/4\mathbb{Z}$.

In the next section, we discuss the problem of finite generation of the module of relations.

Exercises for Section 3.3

EXERCISE 3.14. *In each case of the following matrix A, identify the abelian group which has the given presentation matrix A:*

$$\begin{bmatrix} 2 \\ 1 \end{bmatrix}, \begin{bmatrix} 0 \\ 5 \end{bmatrix}, \begin{bmatrix} 2 & 0 & 0 \end{bmatrix}, \begin{bmatrix} 1 & 0 \\ 1 & 1 \end{bmatrix},$$

$$\begin{bmatrix} 1 & 0 \\ 0 & 1 \\ 0 & 0 \end{bmatrix}, \begin{bmatrix} 2 & 3 \\ 1 & 2 \end{bmatrix}, \begin{bmatrix} 2 & 4 \\ 1 & 2 \end{bmatrix}, \begin{bmatrix} 1 & 2 \\ 3 & 4 \end{bmatrix}$$

4. ACC or Noetherian Condition

In our proof of Theorem 3.6, we needed an additional result, Proposition 3.5 below. Namely, to finish the proof of Theorem 3.6 we must prove Proposition 3.5. For this, we start our discussion with equivalence of two statements:

PROPOSITION 3.4. *Let M be an R-module. The following statements are equivalent:*

a: *Every submodule N of M is finitely generated;*

b: *ACC(Ascending Chain Condition) There is no infinitely strictly increasing chain*

$$N_1 \subset N_2 \subset \cdots$$

of submodules N_i of M.

PROOF. Assume that M satisfies ACC, and let N be a submodule of M. If $N = \{0\}$, then N is generated by 0. If not, then N contains a non-zero element, say v_1. Let N_1 be the submodule of M generated by v_1. Then since $v_1 \in N$, N_1 is automatically a submodule of N. If $N_1 = N$, then we are done. Otherwise, there exists $v_2 \in N - N_1$, and let N_2 be the submodule generated by both v_1 and v_2. If $N_2 = N$, then we are done. Otherwise, there exists $v_3 \in N - N_2$, and let N_2 be the submodule generated by v_1, v_2, and v_3. If $N_3 = N$, then we are done. Otherwise, we repeat this procedure, so that

$$N_1 \subset N_2 \subset N_3 \subset \cdots,$$

but there should exist k such that $N_k = N$ by hypothesis. Thus, N is finitely generated.

Conversely, assume that every submodule of M is finitely generated, and let

$$N_1 \subset N_2 \subset \cdots$$

be an infinite increasing chain of submodules of M. Define U to be the union of these submodules N_i. Then one can easily check that U is a submodule of M. Thus, it is finitely generated by hypothesis. Let $u_1, ..., u_r$ be generators for U. Then each u_i is in one of the modules N_i, and since the chain is increasing, there should exist an integer i such that all the generators are in N_i; hence U is also contained in N_i, and this means that we have

$$U \subset N_i \subset N_{i+1} \subset U.$$

Thus, $U = N_i = N_{i+1} = \cdots$ and hence the chain is not strictly increasing. □

REMARK 3.12. *Here is some historical remark: It is known that Emmy Noether, a Jewish female mathematician whose father was a well-known Algebraic Geometer of* 19th *century, Max Noether, first recognized the importance of the ACC condition above, thus the condition is usually called a "**Noetherian Condition**" in order to celebrate her deep role in developing Modern Algebra of very axiomatic nature. In fact, she is called as the Mother of Modern Algebra, mainly because of her famous Lecture Note, published by Van der Waerden in* 1930*s.*

DEFINITION 3.15. *Any R-module M satisfying one of the equivalent statements of Proposition 3.4 is called a **Noetherian module**.*

LEMMA 3.5. *Let φ: $M \longrightarrow M'$ be a homomorphism of R-modules.*

(1) *If both the kernel and the image of* φ *are finitely generated modules, so is* M.

(2) *If* M *is finitely generated by a generating set* $\{v_1, ..., v_n\}$ *and if* φ *is* surjective, *then* M' *is finitely generated by the images* $\varphi(v_i)$, $i = 1, 2, ..., n$.

PROOF. Let $\{u_1, ..., u_k\}$ be a generating set for the kernel and let

$$\{w_1, ..., w_m\}$$

be a generating set for the image. Let v_i be the preimage of w_i, i.e., an element of M such that $\varphi(v_i) = w_i$. In what follows we prove that $\{u_1, ..., u_k, v_1, ..., v_m\}$ generates M: For this, let $v \in M$ be arbitrary. Then $\varphi(v)$ is a linear combination of $\{w_1, ..., w_m\}$, say $\varphi(v) = a_1w_1 + ... + a_mw_m$. Now consider $v' = a_1v_1 + ... + a_mv_m$. Since $\varphi(v') = \varphi(v)$, we see that $\varphi(v - v') = \varphi(v) - \varphi(v') = 0$; hence $v - v' \in \ker(\varphi)$. Thus, $v - v'$ is a linear combination of elements of $\{u_1, ..., u_k\}$, say $v - v' = b_1u_1 + ... + b_ku_k$. Therefore,

$$\begin{aligned} v &= v' + b_1u_1 + ... + b_ku_k \\ &= a_1v_1 + ... + a_mv_m + b_1u_1 + ... + b_ku_k. \end{aligned}$$

This shows that the set $\{u_1, ..., u_k, v_1, ..., v_m\}$ generates M, and this proves the first assertion.

The proof of the second assertion is very easy. □

COROLLARY 3.7. *Let* N *be a submodule of an* R*-module* M.

(1) *If both* N *and* M/N *are finitely generated, then so is* M.

(2) *If* M *is finitely generated, then so is* M/N. □

We recall that ideals in a commutative ring R, viewed as an R-module, are submodules, and vice versa.

DEFINITION 3.16. *A commutative ring* R *is called* ***Noetherian*** *if every ideal of* R *is finitely generated.*

EXAMPLE 3.14. *Thus, by definition, Principal Ideal Domains, like* $\mathbb{Z}, F[x]$, *where* F *is a field, are Noetherian.*

COROLLARY 3.8. *Let R be a commutative ring which is furthermore* Noetherian. *Then every ideal I of R is contained in a maximal ideal.*

PROOF. If I is not maximal itself, then it is properly contained in a proper ideal I_2, and if I_2 is not maximal, it is properly contained in a proper ideal I_3, and so on. Thus, by the ascending chain condition, the chain $I = I_1 \subset I_2 \subset \cdots$ must be finite; therefore I_k must be maximal for some k. □

REMARK 3.13. *The point of this corollary is that we never need the axiom of choice, as long as we work in a Noetherian ring. In fact, Krull (1899 - 1971) proved that*

THEOREM 3.7. *(Krull) Every non-unit ideal in a commutative ring is contained in a maximal ideal.* □

But as far as we know, its proof requires the application of Zorn's lemma, which is equivalent to the axiom of choice. We can also prove Krull's theorem quite easily, but we decide not to provide its proof here.

Now, we are ready to prove:

PROPOSITION 3.5. *Let M be a finitely generated module over a* Noetherian *ring. Then every submodule of M is finitely generated.*

REMARK 3.14. *We note that we have used this proposition in the proof of Theorem 3.6.*

PROOF. First, suppose that we have proved the proposition in the case of $M = R^m$ for any m. Then for any finitely generated R-module M, we have a canonical surjective homomorphism of R-modules: φ: $R^m \longrightarrow M$ (by Theorem 3.1). Given a submodule N of M, one can easily check that its preimage, $L = \varphi^{-1}(N)$, is a submodule of R^m. Then, L is finitely generated by our assumption. Then since the composition map: $L \longrightarrow N$ is automatically surjective, N is finitely generated by Lemma 3.5.

Thus, we are left to prove the proposition when $M = R^m$. For the proof, we use induction on m.

When $m = 1$, the proposition is true because R is Noetherian and submodules of R and ideals of R are the same.

Now suppose $m > 1$. We then consider the projection

$$\pi\colon R^m \longrightarrow R^{m-1}$$

defined by dropping the last entry: $\pi(a_1, ..., a_m) = (a_1, ..., a_{m-1})$. Then its kernel is $\{(0, ..., 0, a_m)\} \approx R$. Now let N be a submodule of R^m, and let φ: $N \longrightarrow R^{m-1}$ be the restriction of π to N. Then by induction hypothesis its image $\varphi(N)$ is a finitely generated submodule of R^{m-1}. Moreover,

$$\ker(\varphi) = \ker(\pi) \cap N$$

is a submodule of $\ker(\pi) \approx R$; hence it is finitely generated too. Therefore, by Lemma 3.5 N is finitely generated, as required. □

As a corollary of Lemma 3.5, we have:

PROPOSITION 3.6. *Let R be a Noetherian ring, and I be an ideal of R. Then the quotient ring R/I is also Noetherian.*

PROOF. Let $\bar{J}$ be an ideal of R/I, and let $J = \pi^{-1}(\bar{J})$ be the corresponding ideal of R (containing I), where

$$\pi\colon R \longrightarrow R/I$$

is the canonical map. Then by hypothesis, J is finitely generated, say by $(a_1, ..., a_m)$. Then by Lemma 3.5, the finite set $(\bar{a}_1, ..., \bar{a}_m)$ generates $\bar{J}$. □

Having said that, we are going to prove Hilbert Basis Theorem: *Every polynomial ring $R[x]$ over a Noetherian commutative ring is Noetherian,* which is one of his famous theorems.

Let us warm up the situation when R is a field first. The following actually proves that the polynomial ring $F[x]$ over a field F is in fact a Principal Ideal Domain (*PID*):

COROLLARY 3.9. *The polynomial ring $F[x]$ over a field F is Noetherian.*

PROOF. Let I be an ideal in $F[x]$. Since we have the degree function $d\colon F[x] \longrightarrow \mathbb{N} \cup \{0\}$, this allows us to choose a non-zero polynomial $f(x) \in I$ of lowest degree among all the non-zero polynomials in I, say

$$f(x) = a_n x^n + \cdots + a_1 x + a_0, \qquad a_n \neq 0,$$

and let

$$g(x) = b_m x^m + \cdots + b_1 x + b_0, \qquad b_m \neq 0$$

be any non-zero polynomial in I. Then the degree m of $g(x)$ is at least n by our choice of $f(x)$. By induction hypothesis, we assume that all the polynomials in I of degree $< m$ are divisible by $f(x)$. We then note that

$$g(x) - (b_m/a_n)x^{m-n} f(x) = g_1(x) \tag{3.13}$$

is an element of I of degree $< m$, and hence it is divisible by induction hypothesis. Therefore, $g(x)$ is divisible by $f(x)$. □

In the above proof, we note that (3.13) is the first step in the division of $g(x)$ by $f(x)$ with remainder. But, we cannot extent this method directly to an arbitrary ring R because division with remainder requires that the leading coefficient of f be a unit, i.e., an invertible element, of R, or we need to know in advance that a_n divides b_m in the ring R, and this is in general not the case. In order to overcome this difficulty, it turns out that we generally need more generators for I than just one generator.

Here is one of Hilbert's famous theorems:

THEOREM 3.8. *(Hilbert Basis Theorem) If a commutative ring R is Noetherian, then so is the polynomial ring $R[x]$.*

In what follows we are going to present 3 different proofs of Hilbert Basis Theorem, of which the first and second "direct" proofs are quite similar, but the third "indirect" proof by a method of contradiction is probably the shortest proof for his theorem.

For the first proof, we need two Lemmas.

LEMMA 3.6. *Let R be a commutative ring. Then the set L of leading coefficients of all the polynomials in an ideal I of a polynomial ring $R[x]$, together with 0, forms an ideal of R.*

PROOF. If $\alpha \in L$, then it is equal to the leading coefficient a_n of some polynomial $f(x)$ in I; hence for any $r \in R$, $r\alpha$ is clearly the leading coefficient of the polynomial $rf(x)$, which is automatically in I, unless $r\alpha = 0$ by chance. In both cases, $r\alpha \in L$. Now, if $\beta \in L$, then it is equal to the leading coefficient b_m of some polynomial $g(x) \in I$ of degree, say, $m \geq n = \deg(f(x))$. Then since α is also the leading coefficient of $x^{m-n}f(x) \in I$, $\alpha + \beta$ is the leading coefficient of the polynomial

$$h(x) := x^{m-n}f(x) + g(x) \in I,$$

unless $\alpha + \beta = 0$. In both cases, $\alpha + \beta \in L$. This proves that L is an ideal in R. □

LEMMA 3.7. *Let R be a commutative ring, let P_n denote the set of polynomials $f(x)$ in $R[x]$ with $\deg(f(x)) < n$, together with 0, and let $S_n = I \cap P_n$, where I is a polynomial ideal of $R[x]$. Then S_n is an R-submodule of the R-module P_n.*

EXERCISE 3.15. *Prove that this lemma is true.*

THE FIRST PROOF OF THEOREM 3.8. Assume that R is a Noetherian ring, and let I be an ideal of $R[x]$. By Proposition 3.4 it is enough to prove that I is finitely generated. By Lemma 3.6, the set L of leading coefficients of the polynomials in I, together with 0, is an ideal of the Noetherian ring R, so there exists a finite set of generators, say, $(\alpha_1, ..., \alpha_k)$, for I. Then we can choose for each $i, 1 \leq i \leq k$, a polynomial $f_i(x) \in I$ with leading coefficient α_i, and we multiply these polynomials by some powers of x if necessary, in order for all of them to have the same common degree, say, n. We also note that the same α_i is the leading coefficient of the new $f_i(x)$ of the common degree n. (From now on until we finish this proof, this n is fixed.)

We then note: since this n is not in general going to be the smallest degree of polynomials in I, the ideal generated by these f_i's is not in general going to be the ideal I.

Nonetheless, Lemma 3.7 says something: We recall that the R-submodule P_n of $R[x]$ is finitely generated by the monomials

$$1, x, x^2, ..., x^{n-1}.$$

Then since R is Noetherian, we may apply Lemma 3.7 and Proposition 3.5 to conclude that its R-submodule S_n is again finitely generated. Say the R-module S_n has a finite set of generators $\{h_1(x), ..., h_s(x)\}$ with each $h_j \in P_n$; hence each $\deg(h_j(x)) < n$ always. We then have

Claim. $(f_1(x), ..., f_k(x); h_1(x), ..., h_s(x)) = I$, that is, the polynomial ideal I is finitely generated by these polynomials, $f_i(x)$'s and $h_j(x)$'s.

Proof of Claim. Only for simplicity, we denote the ideal

$$(f_1(x), ..., f_k(x); h_1(x), ..., h_s(x))$$

by J. The by our construction, $J \subseteq I$, and hence we only need to prove the opposite inclusion.

For this, let any $g(x) \in I$ be of degree m, and we assume, as an induction hypothesis on m, that all the element of I of degree $< m$ are in the ideal J. If $m < n$ then $g(x) \in I \cap P_n = S_n$; hence $g(x)$ is a linear combination of the $h_j(x)$'s with coefficients in R; consequently $g(x) \in J$. Otherwise, let b be the leading coefficients b_m of the polynomial $g(x)$. Then $b \in L$, the set of all leading coefficients of polynomials in I, so that it is a linear combination of the α_i's, say $b = r_1\alpha_1 + \cdots + r_k\alpha_k$ for some $r_i \in R$. Then since each α_i is the leading coefficient of $f_i(x)$, we see that the polynomial

$$p(x) := x^{m-n} \cdot \sum_{i=1}^{k} r_i f_i(x)$$

has the *same* leading coefficient b and the *same* degree m as $g(x)$; and *moreover* it belongs to the ideal J. Thus,

$$g_1(x) := g(x) - p(x)$$

has degree less than m; hence by induction hypothesis, it is an element of J. Therefore, $g(x) \in J$. $\square$

This finishes the first proof of Hilbert Basis Theorem. $\square$

Here is the second proof of Hilbert Basis Theorem:

THE SECOND PROOF OF THEOREM 3.8. Let R be a Noetherian ring and let I be an ideal of $R[x]$. Then by Proposition 3.4 it is enough to prove that I is finitely generated. For each $n \geq 0$, we define $J_n = \{r \in R : r$ is the leading coefficient of a polynomial in I of degree $n\} \cup \{0\}$. Then it is easy to see that J_n is an ideal of R and that

$$J_n \subset J_{n+1},$$

for each $n \geq 0$, for, for any polynomial $f(x)$, $x \cdot f(x)$ has the same leading coefficient as $f(x)$ does. Since R is Noetherian, there exists N such that $J_N = J_n$ for $n \geq N$. Again by Proposition 3.4, each ideal J_i of R is finitely generated, say,

$$J_i = (r_{i1}, ..., r_{it_i}), \qquad r_{ij} \in R.$$

Now for $i = 1, ..., N$, and $j = 1, ..., t_i$, we can choose, by our construction, a polynomial $f_{ij}(x)$ in I of degree i with leading coefficient r_{ij}. Then we clearly see that the ideal

$$(f_{ij}(x) : 1 \leq i \leq N, 1 \leq j \leq t_i) \subset I,$$

and thus we need to prove the opposite inclusion.

Only for simplicity, we denote the ideal

$$(f_{ij} : 1 \leq i \leq N, 1 \leq j \leq t_i)$$

by I^*. Let any $g(x) \in I$ be of degree n. If $g(x) \equiv 0$, or $n = 0$, then $g(x) \in J_0$; hence it is in I^*. Now for $n > 0$, we assume, as an induction hypothesis on n, that all the elements of I of degree $< n$ are elements of I^*. Let r be the leading coefficient of $g(x)$.

CASE 1. If $n \leq N$, then, since $r \in J_n$, we have

$$r = \sum_{j=1}^{t_n} s_j r_{nj},$$

for some $s_j \in R$. Then the polynomial

$$p(x) := \sum_{j=1}^{t_n} s_j f_{nj}(x)$$

is of degree n, has leading coefficient r, and is an element of I^*. Thus, the polynomial $g(x) - p(x)$ in I has degree $< n$. Thus, by induction hypothesis on n, $g(x) - p(x) \in I^*$; consequently $g(x) \in I^*$. □

CASE 2. Now if $n > N$, then $r \in I_n = I_N$; hence

$$r = \sum_{j=1}^{t_N} s_j r_{Nj},$$

for some $s_j \in R$. Then the polynomial

$$p(x) := \sum_{j=1}^{t_N} s_j x^{n-N} f_{Nj}(x)$$

is of degree n, has leading coefficient r, and is an element of I^*. Thus, the polynomial $g(x) - p(x)$ in I has degree $< n$. Thus, by induction hypothesis on n, $g(x) - p(x) \in I^*$; consequently $g(x) \in I^*$. □

This finishes the second proof of Hilbert Basis Theorem. □

Now, here is the third proof of Hilbert Basis Theorem:

THE THIRD PROOF OF THEOREM 3.8. In what follows, we are going to prove:

If $R[x]$ is not Noetherian, then R is not a Noetherian ring.

Let I be an ideal of $R[x]$ that is not finitely generated. Let $f_1 \in I$ be a polynomial of least degree. If f_k, $k \geq 1$, has already been chosen, let f_{k+1}

be a polynomial of least degree in $I - (f_1, ..., f_k)$. For these polynomials f_k, let n_k be the degree of f_k and $a_k \in R$ be the leading coefficient of f_k, for $k = 1, 2, ...$ Then by our choice of f_k, we see that

$$n_1 \leq n_2 \leq \cdots.$$

Moreover, the increasing chain of ideals in R,

$$(a_1) \subset (a_1, a_2) \subset \cdots$$

must be a strictly increasing chain of ideals in R. For, if $(a_1, ..., a_k) = (a_1, ..., a_{k+1})$, then

$$a_{k+1} = \sum_{i=1}^{k} b_i a_i, \qquad b_i \in R$$

and hence the polynomial

$$p := f_{k+1} - \sum_{i=1}^{k} b_i x^{n_{k+1}-n_i} f_i \in I - (f_1, ..., f_k)$$

is of lower degree than f_{k+1}, contradicting our choice of f_{k+1}, and this finishes the third proof of Hilbert Basis Theorem. □

The Hilbert Basis Theorem shows by induction that

COROLLARY 3.10. *The polynomial ring $R[x_1, ..., x_n]$ in several variables over a Noetherian ring R is Noetherian.* □

Hence the rings $\mathbb{Z}[x_1, ..., x_n]$ and $F[x_1, ..., x_n]$ where F is a field are Noetherian. Also, quotients of Noetherian rings are Noetherian.

Combining Proposition 3.6 with the Hilbert Basis Theorem, we obtain:

COROLLARY 3.11. *Any commutative ring which is a quotient of a polynomial ring over the integers or over a field is Noetherian.* □

Exercises for Section 3.4

EXERCISE 3.16. *Let S be a subset of $\mathbb{C}^n$.*

(1) *Prove that the set $I(S)$ of those polynomials $f(x_1, ..., x_n)$ vanishing on S is an ideal of $\mathbb{C}[x_1, ..., x_n]$.*

(2) *Prove that there is a finite set of polynomials*

$$(f_1(x_1, ..., x_n), ..., f_k(x_1, ..., x_n))$$

such that any polynomial vanishing identically on S is a linear combination of this set, with polynomial coefficients.

EXERCISE 3.17. *Let $V \subset \mathbb{C}^n$ be the locus of zeros of an infinite set of polynomials in n variables:*

$$\begin{cases} f_1(x_1, ..., x_n) & = 0, \\ f_2(x_1, ..., x_n) & = 0, \\ \quad \cdots & \\ \quad \cdots & \end{cases}$$

Prove that there exist a finite subset of these polynomials whose zeros define the same locus.

5. The Structure Theorem for Abelian Groups

In this section we apply what we have developed in the previous sections to the structure theory of abelian groups: *A finitely generated abelian group M is a direct sum of cyclic groups.* To establish it, we are going to apply Theorem 3.3, and others which are its variants and Proposition 3.3. More precisely, since any abelian group is a $\mathbb{Z}$-module, we know by Theorem 3.3 that there exists a diagonal presentation matrix for M, and a proper application of Proposition 3.3 will enable us to interpret the meaning of this diagonal matrix for an abelian group. This means in part that we need to introduce a few more terminologies for modules in order to have a better understanding of the subject.

DEFINITION 3.17. *Let $W_1, ..., W_k$ be submodules of an R-module M.*

(1) *Their **sum** is the submodule which they generate, i.e.,*

$$W_1 + \cdots + W_k := \{w_1 + \cdots + w_k : \text{ each } w_i \in W_i\}.$$

(2) *We say that M is the **direct sum,***

$$M = W_1 \bigoplus \cdots \bigoplus W_k,$$

of submodules W_i of M if

(a) *they generate:* $M = W_1 + \cdots + W_k$;

(b) *they are independent: If* $w_1 + \cdots + w_k = 0$, *with each* $w_i \in W_i$, *then each* $w_i = 0$ *for* $i = 1, ..., k$.

(3) *The* ***direct product*** *of submodules W_i is the product set*

$$W_1 \times \cdots \times W_k = \{(w_1, ..., w_k) : \; w_i \in W_i\}$$

together with vector addition and scalar multiplication:

$$\begin{aligned}(w_1, ..., w_k) + (w_1', ..., w_k') &= (w_1 + w_1', ..., w_k + w_k'),\\ r(w_1, ..., w_k) &= (rw_1, ..., rw_k).\end{aligned}$$

Then it is very easy to check that the direct product $W_1 \times \cdots \times W_k$ is an R-module under these operations. It is also very easy to see that

LEMMA 3.8. *The sum of submodules W_i is a submodule of M and it is the smallest submodule of M among those submodules containing each W_i.* □

It is also very easy to check that

LEMMA 3.9. *M is the direct sum of the submodules W_i if every element $v \in M$ can be written uniquely in the form*

$$v = w_1 + \cdots + w_k,$$

with each $w_i \in W_i$. □

LEMMA 3.10. *Two submodules W_1 and W_2 are independent if and only if $W_1 \cap W_2 = \{0\}$.* □

Finite direct products and finite direct sums are isomorphic, as the following proposition shows:

PROPOSITION 3.7. *Let $W_1, ..., W_k$ be submodules of an R- module M. Then*

a: *The map* σ: $W_1 \times \cdots \times W_k \longrightarrow M$ *defined by sending each* $(w_1, ..., w_k) \mapsto w_1 + \cdots + w_k$ *is a homomorphism of R-modules, and its image is the sum* $W_1 + \cdots + W_k$.

b: *The homomorphism σ is an isomorphism if and only if M is the direct sum of the submodules W_i.*

EXERCISE 3.18. *The proof of this proposition is very easy, so we leave its proof as an exercise.*

THEOREM 3.9. *Let M be a finitely generated abelian group. Then M is a direct sum of finite cyclic subgroups $C_{d_1}, \ldots, C_{d_k}$ and a free abelian group L of finite rank:*

$$M = C_{d_1} \bigoplus \cdots \bigoplus C_{d_k} \bigoplus L,$$

where the order, i.e., the cardinality, d_i of each C_{d_i} is greater than 1, and $d_1 | d_2 | d_3 \ldots .$

REMARK 3.15. *Since we are dealing with subgroups of an abelian group, we will use as usual additive notation for the law of composition in the cyclic group. Namely, C_n denotes a cyclic group generated by a single element v with one relation $nv = 0$. Therefore,*

$$C_n \text{ is obviously isomorphic to } \mathbb{Z}/n\mathbb{Z},$$

where by definition sending the residue of an integer r to rv is the isomorphism: $\mathbb{Z}/n\mathbb{Z} \longrightarrow C_n$.

PROOF. Since $\mathbb{Z}$ is a Noetherian ring and M is a finitely generated $\mathbb{Z}$-module, we can apply the theory of generators and linear relations among them to have a presentation matrix A for M, which is determined by a set of generators and a complete set of relations. Thus, by Proposition 3.3, the matrix A may be replaced by QAP^{-1}, where Q and P are invertible integer matrices. Thus, we may assume that A is diagonal, that the diagonal entries are non-zero, and that each diagonal entry divides the next one. Moreover, thanks to Proposition 3.3 we can drop any column of zeros, and any row and column in which the diagonal entry, i.e., the pivot, is 1. Thus, we may further assume that the diagonal entries d_i of the matrix A are neither 0 nor 1:

$$A = \begin{bmatrix} d_1 & & & & \\ & d_2 & & & \\ & & \cdot & & \\ & & & \cdot & \\ & & & & d_k \\ & & 0 & & \end{bmatrix} \tag{3.14}$$

Thus, we may assume that the size of A is $m \times k$ for some k and m with $k \leq m$. In terms of generators and linear relations, the matrix A means that M is generated by m elements $v_1, ..., v_m$ satisfying a complete set of k relations among these generators:

$$d_1 v_1 = 0, \ d_2 v_2 = 0, \ ..., d_k v_k = 0.$$

Thus, the rest of the generators, $v_{k+1}, ..., v_m$ are free, meaning that we cannot do anything on them because there does not exist any linear relations among them. This is the reason why we need to figure out the complete set of relations among the generators.

For $j = 1, ..., k$, we denote by C_j the cyclic subgroup generated by v_j satisfying $d_j v_j = 0$. We also denote by L the subgroup of M generated by the remaining free generators $v_{k+1}, ..., v_m$. Since they never appear in the set of complete k relations, there is no linear relations among them, and hence L is a free abelian group of rank $m - k$.

To finish the proof, we need to verify that

$$M = C_1 \bigoplus \cdots \bigoplus C_k \bigoplus L,$$

and that each C_j is a cyclic group of order d_j.

First, it is very clear that M is a sum of these subgroups since all the generators v_i's appear either one of them, C_j or L. To show that they are independent, we suppose that we have a relation:

$$z_1 + \cdots + z_k + w = 0,$$

where each $z_j \in C_j$ and $w \in L$. Since each C_j is a cyclic subgroup with generator v_j and since L is a free abelian subgroup of M, we can write each $z_j = r_j v_j$ for some integer $r_j \in \mathbb{Z}$ and each $w = r_{k+1}v_{k+1} + \cdots + r_m v_m$ for some integers $r_i \in \mathbb{Z}$. We thus have a linear relation among the generators $v_1, ..., v_m$:

$$r_1 v_1 + \cdots + r_m v_m = 0.$$

Then since we already know a complete set of relations, the integer column vector $(r_1, ..., r_m)^t$ is a linear combination of the columns of the matrix A; hence for $j \geq k+1$, we see that $r_j = 0$. Thus, $w = 0$. In addition to this, for $j \leq k$, each r_j must be divisible by d_j, say, $r_j = s_j d_j$ for some integer $s_j \in \mathbb{Z}$. Thus, each $z_j = s_j d_j v_j = 0$, and this proves that both the subgroups are independent and the order of each cyclic subgroup C_j is d_j. □

COROLLARY 3.12. *A finite abelian group M is a direct sum of finite cyclic subgroups $C_{d_1}, ..., C_{d_k}$.*

PROOF. Since it is finitely generated and since it is a finite set, we cannot possibly have a subgroup consisting of infinitely many elements, so that by Theorem 3.9 we have

$$M = C_{d_1} \bigoplus \cdots \bigoplus C_{d_k},$$

where $1 < d_1 | d_2 | d_3 ....$ □

It is sometimes convenient to decompose the cyclic groups further into cyclic groups of prime power order. For this we need:

LEMMA 3.11. *Let r, s be two relatively prime integers. Then the cyclic group C_{rs} of order rs is isomorphic to the direct sum of cyclic groups of order r and order s, respectively,*

$$C_{rs} \xrightarrow{\sim} C_r \bigoplus C_s.$$

PROOF. By Proposition 3.7 we already know that $C_r \bigoplus C_s$ is isomorphic to the direct product $C_r \times C_s$. Thus, we are going to establish an isomorphism

$$C_{rs} \xrightarrow{\sim} C_r \times C_s.$$

For this, let v be a generator for C_{rs} and w, z be generators of C_s and C_s, respectively. Then the map sending $kv \mapsto (kw, kz)$ for $0 \le k < rs$, is clearly a Z-module homomorphism with kernel 0. Then since both C_{rs} and $C_r \times C_s$ have the same cardinality rs, they are bijective. □

COROLLARY 3.13. *Moreover, if these cyclic groups C_{rs}, C_r, and C_s, where $(r, s) = 1$, are subgroups of an abelian group M, then $C_{rs} = C_r \bigoplus C_s$.*

PROOF. Since both C_r and C_s are clearly subgroups of C_{rs}, the smallest subgroup $C_r \bigoplus C_s$ containing both C_r and C_s is a subgroup of C_{rs}. But then since they have the same cardinality, they must be the same. □

Combining this result with the Structure Theorem, Theorem 3.9, we get

COROLLARY 3.14. *(Structure Theorem, alternate form) Every finitely generated abelian group is a direct sum of cyclic groups of prime power orders and of a free abelian group of finite rank.* □

THEOREM 3.10. *(Uniqueness for the Structure Theorem)*

- **a:** *Suppose that a finite abelian group M is a direct sum of cyclic groups $C_{d_1} \bigoplus \cdots \bigoplus C_{d_k}$, where $1 < d_1 | d_2 | d_3 \ldots$. Then the integers d_i are (uniquely) determined by the group M.*
- **b:** *The same is true if the decomposition is into prime power orders, that is, if each d_j is the power of a prime.*

EXERCISE 3.19. *We leave the proof of this theorem as an exercise.*

EXAMPLE 3.15. *If the order of a finite abelian group is square free, i.e., a product of distinct primes, then there is no problem since we have just one decomposition. For instance if the order is*

$$30 = 2 \times 3 \times 5,$$

then M must be isomorphic to $C_2 \bigoplus C_3 \bigoplus C_5$. Otherwise, we can have several non-isomorphic decompositions. For instance, if the order is

$$16 = 2^4,$$

then $C_2 \bigoplus C_2 \bigoplus C_4 \xrightarrow{\sim} C_2 \times C_2 \times C_4$ is never isomorphic to $C_4 \bigoplus C_4 \xrightarrow{\sim} C_4 \times C_4$, because the Klein 4 group $C_2 \times C_2$ is never isomorphic to the cyclic group C_4, for instance, and hence the two decompositions are not isomorphic.

REMARK 3.16. *This examples suggests that the counting of elements is sometimes easier if we use Proposition 3.7.*

Now, since a cyclic group C_d of order d is isomorphic to the standard cyclic group $\mathbb{Z}/d\mathbb{Z}$, Proposition 3.7 allows us to express the Structure Theorem as follows:

THEOREM 3.11. *(Product version of the Structure Theorem) Every finitely generated abelian group M is isomorphic to a direct product of cyclic groups*

$$\mathbb{Z}/d_1\mathbb{Z} \times \cdots \times \mathbb{Z}/d_k\mathbb{Z} \times \mathbb{Z}^r,$$

where d_i, r are integers. There is a decomposition in which each d_i divides the next one, and there is another decomposition in which each d_i is a prime power. □

We note that the most closest rings to $\mathbb{Z}$ are Euclidean domains in the sense that both of them are a *PID* (Principal Ideal Domain) and hence Noetherian. For instance,

EXAMPLE 3.16. *The polynomial ring $F[t]$ in one variable t over a field F is a Euclidean domain.*

The classification of abelian groups can carry over to Euclidean domains without essential change: Since Euclidean domain R is Noetherian, any finitely generated R-module M has a presentation matrix in (3.12), and then by the diagonalization theorem (Theorem 3.4) there is a presentation matrix A which is diagonal.

In order to carry along the analogy with abelian groups, we definitely need:

DEFINITION 3.18. *A* ***cyclic R-module*** *M is an R-module which is generated by a single element v.*

COROLLARY 3.15. *Let M be an R-module. Then M is cyclic generated by v if and only if it is isomorphic to a quotient module R/I, where I is the ideal of R consisting of those elements α such that $\alpha v = 0$.*

PROOF. By Theorem 3.1, there exists a surjective homomorphism of R-modules $\varphi\colon R \longrightarrow M$ which sends $r1 \mapsto rv$. Then the ideal I is nothing but the kernel of φ; hence M is isomorphic to R/I. Conversely, if an R-module M is isomorphic to R/I, then the image of the generator 1 for the quotient ring R/I under such an isomorphism will generate M. □

Thus, if R is a Euclidean domain, then the ideal I in this corollary is a principal ideal (α), so that any cyclic module M over a Euclidean domain R is isomorphic to $R/(\alpha)$ for some $\alpha \in R$.

We will need the following theorem in the next section:

THEOREM 3.12.

a: Let M be a finitely generated module over a Euclidean domain R. Then M is a direct sum of cyclic R-modules C_j and a free R-module L. More precisely, there is an isomorphism

$$\varphi\colon M \xrightarrow{\sim} R/(d_1) \times \cdots \times R/(d_k) \times R^r,$$

of M with a direct product of cyclic R-modules $R/(d_i)$ and a free R-module R^r, where r is non-negative, the elements $d_1, ..., d_k$ are not units and not zeros, and $d_i | d_{i+1}$ for each $i = 1, ..., k-1$.

b: The same assertion as part a, except that the condition that d_i divides d_{i+1} is replaced by the following one: Each d_i is a power of a prime element of R. Thus M is isomorphic to a product of the form

$$R/(p_1^{e_1}) \times \cdots \times R/(p_k^{e_k}) \times R^r,$$

where repetition of primes are allowed. □

EXAMPLE 3.17. *Let M be the $F[t]$-module presented by the matrix A of Example 3.7.*

$$A = \begin{bmatrix} (t-2)(t-1) & t-2 \\ (t-1)^3 & (t-2)(t-1) \end{bmatrix}.$$

Then part (1) of Proposition 3.3 tells us that M is also presented by the diagonal matrix

$$A' = \begin{bmatrix} 1 & 0 \\ 0 & (t-2)(t-1)^2 \end{bmatrix}.$$

Again by other parts of Proposition 3.3 we can drop the first row and column from A' because the $(1,1)$ entry of A' has the pivot 1. Thus, M is represented by the 1×1 matrix $[g(t)]$, where

$$g(t) = (t-1)^2(t-2).$$

Therefore, M is a cyclic module, which is isomorphic to $F[t]/(g(t))$. Furthermore, since $g(t)$ is the product of two relatively prime factors, we can further decompose it; consequently it is isomorphic to the direct product of two cyclic modules:

$$M \approx F[t]/(g(x)) \approx F[t]/(t-1)^2 \times F[t]/(t-2).$$

REMARK 3.17. *We note that Principal Ideal Domains (PID's) are in some sense the next closest ring to $\mathbb{Z}$, and hence with slightly more work, we can extend Theorem 3.12 to modules over any PID.*

REMARK 3.18. *Over a PID, it is also true that the prime powers occurring in part b of Theorem 3.12 are unique up to unit factors. A substitute for the counting argument including Proposition 3.7, which proves Theorem 3.10, must be found to prove this fact.*

Exercises for Section 3.5

EXERCISE 3.20. *Determine the number of non-isomorphic abelian groups of order $400 = 2^4 \cdot 5^2$.*

EXERCISE 3.21. *Find an isomorphic direct product of cyclic groups, when M is the abelian group generated by x, y, and z, with the following relations:*

(1) $3x + 2y + 8z = 0, 2x + 4z = 0.$
(2) $x + y = 0, 2x = 0, 4x + 2z = 0, 4x + 2y + 2z = 0.$
(3) $2x + y = 0, x - y = 3z = 0.$
(4) $2x - 4y = 0, 2x + 2y + z = 0.$

EXERCISE 3.22. *Find a direct sum of cyclic groups which is isomorphic to the abelian group presented by the matrix*

$$A = \begin{bmatrix} 3 & 3 & 3 \\ 3 & 3 & 0 \\ 3 & 0 & 3 \end{bmatrix}.$$

EXERCISE 3.23. *Let R be a ring, and let M be an R-module, presented by a diagonal $m \times n$ matrix A:*

$$R^m / AR^n \approx M.$$

Let $(v_1, ..., v_m)$ be the corresponding generators of M, and let d_i be the diagonal entries of A. Prove that M is isomorphic to a direct product of the modules $R/(d_i)$.

6. $F[t]$-Modules and Endomorphisms on a Vector Space

We have so far in this chapter extended the method developed first for abelian groups to modules over Euclidean domains. As we shall see soon, it turns out that a theory on an $F[t]$-module can be equivalently described as a theory on a vector space with an endomorphism, and conversely. But, we first develop the theory without talking about $F[t]$-modules and this will provide the reader with a chance to see how polynomials in one variable can play its role in our study of endomorphisms on a vector space.

6.1. The Minimal Polynomial of an Endomorphism. We recall that similar matrices, PAP^{-1} with invertible matrices P, have the same determinant and the same trace as the square matrix A. These invariants take values in the field of scalars and hence they are sometimes called "**numerical invariants**". A different kind of invariant for similarity is the function $F^{n^2} \longrightarrow F[\lambda]$ which associates to each $n \times n$ matrix A its characteristic polynomial, $\det(\lambda I - A)$, in one unknown λ over a field F. In this subsection, we are going to construct another such "polynomial invariant", called the "*minimal polynomial*" of A.

DEFINITION 3.19. *We denote by*

$$End(V)$$

the set of all endomorphisms on a vector space V over a field F.

We then easily see that for each scalar α, left multiplication by α is certainly an endomorphism on V, and we denote this endomorphism by

$$\alpha I \colon V \longrightarrow V.$$

Thus, $End(V)$ contains the collection F of scalars as a subset. If $t \colon V \longrightarrow V$ is an endomorphism on V, then its scalar multiple,

$$\alpha t \colon V \longrightarrow V,$$

is again an endomorphism on V, and hence any polynomial expression in t of positive degree is again an endomorphism on V.

In fact, we have

LEMMA 3.12. *Let V be a vector space over a field F.*

- **a:** *$End(V)$ is an F-vector space under pointwise addition and pointwise scalar multiplication.*
- **b:** *$End(V)$ is a ring (usually non-commutative) under the operations of pointwise sum, composition, and selection of the identity map on V as a unit.*
- **c:** *The subring of $End(V)$ generated by an endomorphism*

$$t : V \longrightarrow V$$

and scalar multiples of the identity map I on V is a commutative ring containing F and t.

d: *If V has dimension n, then there exists an isomorphism*

$$End(V) = Hom(F^n, F^n) \approx F^{n\times n}$$

of F-vector spaces.

PROOF. It is very routine to verify them, so we leave their proofs as an exercise. □

In particular, part c of this lemma means that, by the universally repelling property of the polynomial ring, there exists a unique ring homomorphism from the polynomial ring:

$$E_t\colon F[x] \longrightarrow End(V)$$

which sends the indeterminate x to the generator t and whose restriction on F is the identity map.

THEOREM 3.13. *Let $t\colon V \longrightarrow V$ be an endomorphism on a finite dimensional vector space V over F. Then there exists (unique) monic (not-necessarily irreducible) polynomial $m(x)$ of least degree such that $m(t) = 0$ under*

$$E_t : F[x] \longrightarrow End(V).$$

PROOF. Since $\dim(V) = n$, $End(V) = Hom(F^n, F^n)$ and hence it is isomorphic to the vector space $F^{n\times n}$ of dimension n^2. Thus, for example, the $n^2 + 1$ of endomorphisms on V

$$1, t, ..., t^{n^2}$$

is linearly dependent. This just means that there exists a non-constant polynomial $f(x) \in F[x]$ such that $f(t) = 0$. That is, the kernel of the evaluation homomorphism E_t is bigger than (0), hence since $F[x]$ is a *PID*, the kernel is generated by a single (not-necessarily irreducible) polynomial $m(x)$ over F. If we require this polynomial $m(x)$ to be monic, then it is uniquely determined. □

DEFINITION 3.20. *This not-necessarily irreducible polynomial $m(x)$ over F is called the* ***minimal polynomial*** *of $t\colon V \longrightarrow V$.*

COROLLARY 3.16. *Let V be a finite-dimensional vector space over a field F. Then the minimal polynomial of an endomorphism $t \in End(V)$ is identical to the minimal polynomial of any matrix representing t.*

PROOF. Since the subring of the matrix ring $F^{n\times n}$ generated by left multiplication by an $n \times n$ matrix A: $F^n \longrightarrow F^n$ and scalar multiples of the identity map I on F^n is a commutative ring containing F and A, we have exactly the same theorem for the ring $F^{n\times n}$ of square matrices as Theorem 3.13, and hence we have the minimal polynomial of A, and then since for each basis $\mathbf{B}$ of V the function:

$$End(V) \longrightarrow F^{n\times n}$$

sending each endomorphism $t \in End(V)$ to the matrix A of t with respect to $\mathbf{B}$ is an isomorphism of rings, we have the desired result. □

COROLLARY 3.17. *Similar matrices have the same minimal polynomial.*

PROOF. This is because similar matrices represent the same endomorphism relative to different bases. □

6.1.1. ***Nilpotent Endomorphisms.***

DEFINITION 3.21. *An endomorphism t: $V \longrightarrow V$ is called* **nilpotent** *if there exists positive integer k such that $t^k = 0$.*

Thus, the minimal polynomial of a nilpotent endomorphism t is

$$x^n$$

for some n.

In particular if $\dim(V) = n =$ the degree of the minimal polynomial of a nilpotent endomorphism t on V, then there exists a vector $w \in V$ such that $t^{n-1}(w) \neq 0$; then the n vectors of V,

$$w, t(w), ..., t^{n-1}(w),$$

are linearly independent. For, if $0 = c_0w + c_1t(w) + \cdots + c_{n-1}t^{n-1}(w)$, say, with the first non-zero coefficient c_j, then $0 = t^{n-j-1}(c_jt^j(w)) = c_jt^{n-1}(w)$, which is a contradiction to $t^{n-1}(w) \neq 0$. Thus, if we take them as a basis for V, we have

$$t(w, t(w), ..., t^{n-1}(w)) = (w, t(w), ..., t^{n-1}(w))A,$$

where

$$A = \begin{bmatrix} 0 & 0 & .. & 0 & 0 \\ 1 & 0 & .. & 0 & 0 \\ 0 & 1 & .. & 0 & 0 \\ . & . & & . & . \\ & & .. & & \\ . & . & & . & . \\ 0 & 0 & .. & 0 & 0 \\ 0 & 0 & .. & 1 & 0 \end{bmatrix}.$$

This $n \times n$ matrix surely has the minimal polynomial x^n; and moreover we can easily see that its characteristic polynomial is also x^n, which is a special case of the fact that the minimal polynomial of t divides the characteristic polynomial of t.

6.1.2. ***Companion Matrix.*** Now suppose we are given a monic polynomial of degree m,

$$g(x) = g_0 + g_1 x + \cdots g_{m-1} x^{m-1} + x^m.$$

From the above arguments for nilpotent endomorphisms, we can easily construct an $m \times m$ matrix:

$$A_g = \begin{bmatrix} 0 & 0 & .. & 0 & -g_0 \\ 1 & 0 & .. & 0 & -g_1 \\ 0 & 1 & .. & 0 & . \\ . & . & & . & . \\ & & .. & & \\ . & . & & . & . \\ 0 & 0 & .. & 0 & -g_{m-2} \\ 0 & 0 & .. & 1 & -g_{m-1} \end{bmatrix}.$$

DEFINITION 3.22. A_g *is called the* ***companion matrix*** *of the monic polynomial,*

$$g(x) = g_0 + g_1 x + \cdots g_{m-1} x^{m-1} + x^m$$

in $F[x]$.

PROPOSITION 3.8. *The companion matrix* A_g *of a monic polynomial*

$$g(x) = g_0 + g_1 x + \cdots g_{m-1} x^{m-1} + x^m$$

of degree m *is an* $m \times m$ *matrix with minimal polynomial* $g(x)$ *and characteristic polynomial* $(-1)^m g(x)$.

PROOF. We note that left multiplication by A_g is a linear transformation on the spaceF^m, $t : F^m \longrightarrow F^m$, and it sends the standard basis elements ε_j of F^m as follows:

$$\begin{aligned} t(\varepsilon_j) &= \varepsilon_{j+1} \qquad \text{for } 1 \leq j \leq m-1, \\ t(\varepsilon_m) &= -\varepsilon_1 g_0 - \varepsilon_2 g_1 - \cdots - \varepsilon_m g_{m-1}. \end{aligned}$$

Thus, by the first $m-1$ equations, we see that

$$\begin{aligned} \varepsilon_m &= t(\varepsilon_{m-1}) = t^2(\varepsilon_{m-2)} = \ldots = t^{m-1}(\varepsilon_1), \\ \varepsilon_{j+1} &= t^j(\varepsilon_1), \end{aligned}$$

for each $j < m$. Thus, we get

$$t(\varepsilon_m) = t^m(\varepsilon_1),$$

and hence replacing each ε_{j+1} in the second equation with this, we get

$$t^m(\varepsilon_1) = -(g_0 + g_1 t + \cdots + g_{m-1} t^{m-1})(\varepsilon_1).$$

Thus, we see that

$$g(t)(\varepsilon_1) = 0,$$

and consequently we also see that

$$g(t)(\varepsilon_{j+1}) = g(t)(t^j(\varepsilon_1)) = t^j(g(t)(\varepsilon_1)) = 0,$$

for every $j = 1, 2, \ldots, m-1$. Therefore, $g(t)(w) = 0$ for every vector w in F^m; hence we must have

$$g(t) = 0.$$

On the other hand, if $f(x) = \sum_{j=1}^{d} f_j x^j$ is a polynomial of lower degree $d < m$, then, in a similar manner, we get $\left(\sum_{j=1}^{d} f_j t^j\right)(\varepsilon_1) = \dots = \sum_{j=1}^{d} f_j \varepsilon_{j+1}$, which is a linear combination of the ε_j's; hence it is zero only when every coefficient f_j of $f(x)$ is zero, because the ε_j are the standard basis of F^m. Therefore, $g(x)$ is indeed the minimal polynomial for t and hence for A_g.

Then one can easily check that

$$\begin{aligned} \det(A_g - \lambda I) &= \det \begin{bmatrix} -\lambda & 0 & .. & 0 & -g_0 \\ 1 & -\lambda & .. & 0 & -g_1 \\ 0 & 1 & .. & 0 & . \\ . & . & & . & . \\ & & .. & & \\ . & . & & . & . \\ 0 & 0 & .. & -\lambda & -g_{m-2} \\ 0 & 0 & .. & 1 & -\lambda - g_{m-1} \end{bmatrix} \\ &= (-1)^m g(\lambda), \end{aligned}$$

where we get the last equality by the minor expansion of it through the last column. □

EXAMPLE 3.18. *When $m = 4$, we have*

$$\begin{aligned} \det(A_g - \lambda I) &= \det \begin{bmatrix} -\lambda & 0 & 0 & -g_0 \\ 1 & -\lambda & 0 & -g_1 \\ 0 & 1 & -\lambda & -g_2 \\ 0 & 0 & 1 & -\lambda - g_3 \end{bmatrix} \\ &= (-1)^4(g_0 + g_1\lambda + g_2\lambda^2 + g_3\lambda^3 + \lambda^4) \\ &= (-1)^4 g(\lambda). \end{aligned}$$

This matrix is special in the following sense: Let $t \in End(V)$ be a non-trivial endomorphism on a vector space of dimension n. Then we can easily see that there exists a non-zero vector $w \in V$ such that $t(w) \neq 0$, thus we can have a sequence of vectors: $w, t(w), t^2(w), \dots.$ in V. Let

$$C_w(t)$$

be the subspace of V spanned by these vectors, $w, t(w), t^2(w), \ldots$ We can easily see that

LEMMA 3.13. *For a non-zero vector $w \in V$ of dimension n, $C_w(t)$ is t-invariant, i.e., $t(C_w(t)) \subset C_w(t)$.* $\square$

DEFINITION 3.23. *$C_w(t)$ is called the t-**cyclic subspace** of V.*

Moreover, we have

LEMMA 3.14. *For a non-zero vector $w \in V$ of dimension n, there exist $m \leq n = \dim(V)$ and scalars $g_0, g_1, \ldots, g_{m-1} \in F$ such that*

$$\{w, t(w), \ldots, t^{m-1}(w)\}$$

is a basis of the t-cyclic subspace $C_w(t)$ of dimension m, and

$$t(w, t(w), \ldots, t^{m-1}(w)) = (w, t(w), \ldots, t^{m-1}(w))A_g,$$

where

$$g = g_0 + g_1 x + \ldots + g_{m-1}x^{m-1} + x^m$$

is a monic polynomial in x of degree m, and hence

$$A_g = \begin{bmatrix} 0 & 0 & .. & 0 & -g_0 \\ 1 & 0 & .. & 0 & -g_1 \\ 0 & 1 & .. & 0 & . \\ . & . & .. & . & . \\ . & . & & . & . \\ 0 & 0 & .. & 0 & -g_{m-2} \\ 0 & 0 & .. & 1 & -g_{m-1} \end{bmatrix}$$

is the matrix of $t|_{C_w(t)}$ with respect to this basis.

PROOF. We know that $w \neq 0$ itself is linearly independent and since V has dimension n,there exists an integer m such that

$$w, t(w), \ldots, t^{m-1}(w)$$

are linearly independent, but

$$w, t(w), \ldots, t^{m-1}(w), t^m(w)$$

are linearly dependent with $m \leq n$. Thus, we have a linear dependence relation:

$$g_0 w + g_1 t(w) + ... + g_{m-1} t^{m-1}(w) + t^m(w) = 0,$$

for some scalars $g_0, ..., g_{m-1} \in F$. That is,

$$(g_0 + g_1 t + ... + g_{m-1} t^{m-1} + t^m)(w) = 0.$$

Thus, we obtain the matrix A_g of $t_{C_w(t)}$ with respect to this basis of $C_w(t)$, where $g = g_0 + g_1 x + ... + g_{m-1} x^{m-1} + x^m$. □

Moreover, we see that

LEMMA 3.15. *The polynomial $g(x)$ divides another polynomial $f(x)$ if and only if $f(t)(w) = 0$.*

PROOF. If $f(t)(w) = 0$, then we apply the Euclidean algorithm:

$$f(x) = q(x)g(x) + r(x),$$

with $0 \leq \deg(r(x)) < \deg(g(x))$ for unique $q(x)$ and $r(x) = b_0 + b_1 x + ... + b_{m-1} x^{m-1}$. Thus, $f(t)(w) = 0$ implies that $r(t)(w) = 0$, but then since $w, t(w), ..., t^{m-1}(w)$ is linearly independent, each coefficient $b_i = 0$. Thus, $g(x)$ should divide $f(x)$. The converse of the lemma is trivial. □

We thus proved

PROPOSITION 3.9. *Let $t \in End(V)$ be an endomorphism on V of dimension n. Then for any non-zero vector $w \in V$, there exists a monic (not-necessarily irreducible) polynomial $g(x) \in F[x]$ of least degree such that $g(t)(w) = 0$.* □

DEFINITION 3.24. *Such a not-necessarily irreducible polynomial $g(x)$ is called t-**minimum polynomial** of the non-zero vector w.*

LEMMA 3.16. *For the restriction of t on t-invariant subspace $C_w(t)$, $t|_{C_w(t)} \in End(C_w(t))$, g is the characteristic polynomial of $t|_{C_w(t)}$.*

PROOF. Write $f(x)$ be the characteristic polynomial of the matrix A_g of $t|_{C_w(t)}$. Adding x^i multiple of each i-th row to the first row for

$2 \leq i \leq n$, we obtain

$$\begin{aligned} f(x) &= \det \begin{bmatrix} x & 0 & \cdots & 0 & g_0 \\ -1 & x & \cdots & 0 & g_1 \\ 0 & -1 & \cdots & 0 & g_2 \\ \vdots & \vdots & & \vdots & \vdots \\ 0 & 0 & \cdots & x & g_{m-2} \\ 0 & 0 & \cdots & -1 & x+g_{m-1} \end{bmatrix} \\ &= \det \begin{bmatrix} 0 & 0 & \cdots & 0 & g(x) \\ -1 & x & \cdots & 0 & g_1 \\ 0 & -1 & \cdots & 0 & g_2 \\ \vdots & \vdots & & \vdots & \vdots \\ 0 & 0 & \cdots & x & g_{m-2} \\ 0 & 0 & \cdots & -1 & x+g_{m-1} \end{bmatrix}. \end{aligned}$$

Thus, the minor expansion through the first row gives:

$$f(x) = (-1)^{m+1} g(x)(-1)^{m-1} = g(x).$$

□

PROPOSITION 3.10. *Let t be an endomorphism on a vector space V of dimension n, W its t-invariant subspace. Then the characteristic polynomial of $t|_W$ divides that of t.*

PROOF. Let $\mathbf{C} = (u_1, ..., u_r)$ be a basis of W and we extend it to a basis of V:

$$\mathbf{B} = (u_1, ..., u_r; v_1, ..., v_s).$$

Then the matrix of t with respect to $\mathbf{B}$ looks like:

$$t((u_1, ..., u_r; v_1, ..., v_s)) = (u_1, ..., u_r; v_1, ..., v_s)M,$$

$$M = \begin{bmatrix} A & B \\ & C \end{bmatrix},$$

where A is the matrix of t_W with respect to $\mathbf{C} = (u_1, ..., u_r)$. Thus, the characteristic polynomial $\det(xI - M)$ of t is expanded as follows:

$$\begin{aligned} \det(xI - M) &= \det \begin{bmatrix} xI_r - A & -B \\ 0 & xI_s - C \end{bmatrix} \\ &= \det(xI_r - A) \cdot \det(xI_s - C). \end{aligned}$$

□

COROLLARY 3.18. *The characteristic polynomial of* $t|_{C_w(t)}$, *i.e.,* $g(x)$ *constructed in Lemma 3.14, divides the characteristic polynomial of* $t \in End(V)$. □

THEOREM 3.14. *(Caley-Hamilton Theorem) Let* t *be an endomorphism on a vector space* V *of dimension* n *over a field* F *and let* $f(x)$ *be its characteristic polynomial in* $F[x]$. *Then* $f(t)$ *is the zero endomorphism on* V.

PROOF. We already know that $f(t)$ is an endomorphism on V, so

$$f(t)(0) = 0.$$

Now for a non-zero vector $w \in V$, we have seen that $C_w(t)$ is a t-invariant subspace of V with a basis $(w, t(w), ..., t^{m-1}(w))$ and the matrix of $t|_{C_w(t)}$ with respect to this basis is the companion matrix

$$A_g = \begin{bmatrix} 0 & 0 & \cdots & 0 & -g_0 \\ 1 & 0 & \cdots & 0 & -g_1 \\ 0 & 1 & \cdots & 0 & -g_2 \\ \vdots & \vdots & & \vdots & \vdots \\ 0 & 0 & \cdots & 0 & -g_{m-2} \\ 0 & 0 & \cdots & 1 & -g_{m-1} \end{bmatrix},$$

where $g(x) = g_0 + g_1 x + \cdots g_{m-1} x^{m-1} + x^m$ is the minimal polynomial of $t|_{C_w(t)}$ and furthermore, we also have seen that $g(x)$ characteristic polynomial of $t_{C_w(t)}$. Then by the above corollary, $f(x) = h(x) \cdot g(x)$ for some

$h(x) \in F[x]$. Thus, for any non-zero vector $w \in V$, $g(t)(w) = 0$ implies

$$f(t)(w) = 0.$$

□

COROLLARY 3.19. *(Matrix Version of Caley-Hamilton Theorem) Every square matrix A satisfies its own characteristic equation, i.e., if $\varphi(x) = \det(xI - A)$. then we have*

$$\varphi(A) = 0.$$

PROOF. Since we have an isomorphism

$$Mat_n(F) \approx End(F^n),$$

of F-vector spaces, each matrix A is an endomorphism on F^n, left multiplication by A. Thus, by the above theorem $f(A)$ is the zero endomorphism on F^n and hence it is the zero matrix, as claimed. □

Here is the second proof of the matrix version of Caley-Hamilton Theorem, which is more elementary than the above proof because we do not have to interpret a matrix as left multiplication on a finite dimensional vector space V.

THE SECOND PROOF OF COROLLARY 3.19. Write

$$\det(xI - A) = c_0 + \cdots + c_{n-1}x^{n-1} + x^n,$$

where $c_i \in F$. For our purpose, we denote this polynomial in x by

$$\varphi(x).$$

Thus,

$$\varphi(A) = \sum_{i=0}^{n} c_i A^i, \qquad c_n = 1,$$

is a polynomial in the matrix A. Then since we have

$$x^i I - A^i = (x^{i-1}I + \cdots + +xA^{i-2} + A^{i-1})(xI - A),$$

we can express A^i as

$$A^i = x^i I - (x^{i-1}I + \cdots + +xA^{i-2} + A^{i-1})(xI - A).$$

Thus,

$$\begin{aligned}\varphi(A) &= \sum_{i=0}^{n} c_i x^i I - \sum_{i=0}^{n} c_i (x^{i-1}I + \cdots + +xA^{i-2} + A^{i-1})(xI - A) \\ &= c(x)I - \sum_{i=0}^{n} c_i (x^{i-1}I + \cdots + +xA^{i-2} + A^{i-1})(xI - A),\end{aligned}$$

where $c(x)I = \det(xI - A)I$. On the other hand, we may recall that

$$c(x)I = adj(xI - A) \cdot (xI - A).$$

Thus, we have

$$\begin{aligned}\varphi(A) &= adj(xI - A) \cdot (xI - A) \\ &\quad - \sum_{i=0}^{n} c_i (x^{i-1}I + \cdots + A^{i-1})(xI - A) \\ &= \left\{ adj(xI - A) - \sum_{i=0}^{n} c_i (x^{i-1}I + \cdots + A^{i-1}) \right\} \cdot (xI - A) \\ &= D \cdot (xI - A),\end{aligned}$$

where D is a matrix with entries in $F[x]$. Then we note that the matrix $\varphi(A)$ on the left side of this equality does not contain the unknown x, whereas since the right side is the product of two matrices with entries in $F[x]$ where $xI - A$ is not the zero matrix, D must be the zero matrix. Otherwise, we write D as

$$D = D_k x^k + \ldots + D_1 x + D_0,$$

where D_k is not the zero matrix and each D_j, $0 \leq j \leq k$, is a matrix with entries only in F. Then the product, $D \cdot (xI - A)$ when expanded, has the non-zero leading term $D_k x^{k+1}$ in x, but this is impossible because $\varphi(A)$ on the left side is a matrix of constants (degree 0 in x). □

COROLLARY 3.20. *The minimal polynomial of any $n \times n$ matrix divides the characteristic polynomial of A, and hence has degree at most n.*

PROOF. This is true by Theorem 3.13. □

REMARK 3.19. *Theorem 3.15 and Remark 3.21 below completely describe how we can obtain the minimal polynomial of an endomorphism on a finite-dimensional vector space V over a field F.*

6.2. A One to One Correspondence Between $F[t]$-modules and Endomorphisms on a Vector space V. Suppose that we are given an endomorphism on a vector space V, $T\colon V \longrightarrow V$ on a vector space V over a field F. We can use this endomorphism to make V into a module over the polynomial ring $F[t]$. For this

DEFINITION 3.25. *For a given polynomial $f(t) = a_n t^n + \cdots + a_1 t + a_0 \in F[t]$ and for a given vector $v \in V$, we define scalar multiplication:*

$$f(t).v = a_n T^n(v) + a_{n-1} T^{n-1}(v) + \cdots + a_1 T(v) + a_0 v.$$

We note that it is a generalization of left multiplication by a scalar $\alpha \in F$,

$$\alpha.v = \alpha v.$$

Then we can easily verify that

LEMMA 3.17. *Under this extended scalar multiplication, the module V over a field F becomes a module over the polynomial ring $F[t]$ in one unknown t.* □

Conversely, let V be an $F[t]$-module. Then by restricting scalars, we can easily check that V is a module over F. Moreover, we easily see that

LEMMA 3.18. *Left multiplication by the scalar $t \in F[t]$,*

$$t.v = tv$$

is an endomorphism T on V, $T\colon V \longrightarrow V$ sending $v \mapsto T(v) = tv$, and thus, left multiplication by a polynomial $f(t) = a_n t^n + \cdots + a_1 t + a_0 \in F[t]$,

$$\begin{aligned} f(t).v &= a_n t^n.v + a_{n-1}t^{n-1}.v \cdots + a_1 t.v + a_0.v \\ &= a_n t^{n-1}(T(v)) + a_{n-1}t^{n-2}(T(v)) \cdots + a_1 T(v) + a_0 v \\ &= a_n T^n(v) + \cdots + a_1 T(v) + a_0 v \end{aligned}$$

is an endomorphism on V. □

This establishes

PROPOSITION 3.11. *There exists a one to one correspondence between the set of all endomorphisms on a vector space V, $T\colon V \longrightarrow V$ over a field F, and the set of all modules V over the polynomial ring $F[t]$ in one unknown t.* □

That is,

REMARK 3.20. *An endomorphism on an F-vector space and an $F[t]$-module are equivalent concepts.*

6.3. Rational Canonical Form. Now if $\dim_F(V) = n$, then we can apply the theory developed in the last section of this Chapter: Given an endomorphism T on a finite-dimensional vector space V of dimension n over a field F, we view V as a *finitely generated* $F[t]$-module, and hence it has a *presentation matrix* A.

PROBLEM 1. *Think about how we are going to obtain a presentation matrix A of the finitely generated $F[t]$-module V from an endomorphism $t\colon V \longrightarrow V$. As the proof of the following theorem shows we are only able to obtain it very indirectly.*

THEOREM 3.15. *Let $t\colon V \longrightarrow V$ be an endomorphism of an n dimensional F-vector space V, and let $M = (M_{ij})$ be the matrix of t with respect to a basis $(v_1, \ldots, v_n)$ of V. Then the matrix*

$$M - tI \in F[t]$$

of polynomials in $F[t]$ represents the finitely generated $F[t]$-module V. That is, the matrix $M - tI$ presents the $F[t]$-module V.

PROOF. Since $(v_1, ..., v_n)$ still generates the $F[t]$-module V over $F[t]$, we have the canonical surjection

$$p\colon (F[t])^n \longrightarrow V.$$

Let $(b_1, ..., b_n)$ be a basis of the free $F[t]$-module $(F[t])^n$. By definition (cf. Theorem 3.1), each b_i is mapped to v_i under this p, that is,

$$p(b_i) = v_i, \qquad \text{for } i = 1, ..., n.$$

Then since F is a field it is a Noetherian ring, and hence $F[t]$ is a Noetherian ring, and hence $(F[t])^n$ is a Noetherian module by Proposition 3.5, and hence the kernel $\ker(p)$ is a finitely generated $F[t]$-module.

Claim. $\ker(p)$ is generated by n elements u_j of $(F[t])^n$,

$$u_j := b_j t - \sum_{i=1}^{n} b_i M_{ij} \in (F[t])^n, \qquad j = 1, 2, ..., n.$$

PROOF OF CLAIM. We note that $t(v_j) = \sum_{i=1}^{n} v_i M_{ij}$. Thus, for each $j = 1, 2, ..., n$, we have

$$\begin{aligned} p(b_j t - \sum_{i=1}^{n} b_i M_{ij}) &= p(b_j)t - \sum_{i=1}^{n} p(b_i) M_{ij} = v_j t - \sum_{i=1}^{n} v_i M_{ij} \\ &= t(v_j) - \sum_{i=1}^{n} v_i M_{ij} = 0. \end{aligned}$$

Thus, the submodule S of $(F[t])^n$ generated by these n elements $b_j t - \sum_{i=1}^{n} b_i M_{ij}$ over $F[t]$ is contained in $\ker(p)$. That is, $S \subseteq \ker(p)$. For the reverse inclusion, let

$$u := \sum_{j=1}^{n} b_j f_j(t)$$

be an element of $\ker(p)$. Then since each of its component $b_j t^k$ can be recursively written as

$$\begin{aligned} b_j t^k &= b_j t \cdot t^{k-1} = (u_j + \sum_{i=1}^{n} b_i M_{ij}) \cdot t^{k-1} \\ &= u_j t^{k-1} + \sum_{i=1}^{n} b_i t^{k-1} M_{ij} \\ &= u_j t^{k-1} + \sum_{i=1}^{n} (u_i t^{k-2} + \sum_{\alpha=1}^{n} b_\alpha t^{k-2} M_{\alpha j}) M_{ij} \\ &= \ldots\ldots. \\ &= u'_j + \sum_{i=1}^{n} b_i \gamma_{ij}, \end{aligned}$$

with $u'_j \in \ker(p)$ and $\gamma_{ij} \in F$, we can write u as an F-linear combination $\sum_{i=1}^{n} b_i \gamma_i$ of b_i's modulo S. Then since V is a vector space over F with basis $(v_1, v_2, ..., v_n)$ by hypothesis, $p(u) = 0$ forces all the scalars γ_i of b_i must be zero. Thus, $u \in S$. □

Thus, we *define* an $F[t]$-module homomorphism

$$s : (F[t])^n \longrightarrow (F[t])^n$$

by sending

$$b_j \longmapsto \sum_{i=1}^{n} b_i M_{ij} - b_j t = \sum_{i=1}^{n} b_i (M_{ij} - \delta_{ij} t), \qquad j = 1, 2, ..., n,$$

in order to have $img(s) = \ker(p)$. Moreover, this assignment explicitly tells us that the matrix of s with respect to the basis $(b_1, ..., b_n)$ of $(F[t])^n$ is $M - tI$. Therefore, we have

$$(F[t])^n / \ker(p) = (F[t])^n / img(s) = (F[t])^n / (M - tI)(F[t])^n$$

is the *cokernel* of s, and consequently $M - tI$ is a presentation matrix of the finitely generated $F[t]$-module V. □

REMARK 3.21. *Thus, applying the procedures explained in detail both in the proof of Theorem 3.9 and Proposition 3.3 to the matrix*

$$M - tI$$

we then obtain a "reduced" presentation matrix A of the $F[t]$-module V.

We recall that if the presentation matrix (with polynomial entries) A of an endomorphism $T\colon V \longrightarrow V$ is of size $r \times s$, then r is the number of chosen generators for the module V and s is the number of linear relations among these generators.

Regarding V as an $F[t]$-module, we can apply Theorem 3.12 to conclude that V is a direct sum of cyclic $F[t]$-submodules W_i,

$$V = W_1 \bigoplus \cdots \bigoplus W_k,$$

where each W_i is isomorphic to $F[t]/(p_i^{e_i})$, each $p_i(t)$ being an irreducible polynomial in $F[t]$. Since we are assuming that V is a finite-dimensional vector space, we have no free summand in this expression.

DEFINITION 3.26. *Such a module is called a* ***torsion module of finite type****.*

In fact, we have seen that Theorem 3.15 and Remark 3.21 tell us how to compute each invariant factor, $p_i^{e_i}$, from the matrix $M - tI$, where M is the matrix of the endomorphism $T\colon V \longrightarrow V$.

DEFINITION 3.27. *The last polynomial $p_k^{e_k}$ (not in part b but) in part a of Theorem 3.12 is called the* ***minimal polynomial*** *of T. Or equivalently, the least common multiple of those invariant factors $p_i^{e_i}$ in part b of Theorem 3.12 is called the* ***minimal polynomial*** *of T.*

Now, we definitely need to translate such a direct sum into an information on an endomorphism $T\colon V \longrightarrow V$ defined by

$$T(v) = tv.$$

For this, we first recall Corollary 3.15 which says that a cyclic $F[t]$-module W is isomorphic to the quotient $F[t]$-module

$$F[t]/(f(t)),$$

where

$$f(t) = t^m + a_{m-1}t^{m-1} + \cdots + a_1 t + a_0$$

is a monic polynomial in $F[t]$. This is because $F[t]$ is a *PID* (principal ideal domain). Thus, if we denote w_0 be the residue of $1 \in F[t]$ in $F[t]/(f(t))$, i.e., $w_0 = 1 + (f(t)) \in F[t]/(f(t))$, we have

$$f(t)w_0 = 0$$

in $F[t]/(f(t))$. And by this construction, we see that $f(t)$ generates the module of relations, that is, the set $\{f(t)\}$ is a complete set of relations.

On the other hand, we note that the elements $w_0, tw_0, ..., t^{m-1}w_0$ form an F-basis for $F[t]/(f(t))$, because the set $\{t^i : i = 0, 1, 2, ...\}$ is an F-basis of the F-vector space $F[t]$. If we denote this basis by $w_i = t^i w_0$, we then have

$$tw_0 = w_1, tw_1 = w_2, ..., tw_{m-2} = w_{m-1}.$$

In addition to this, since we have

$$\begin{aligned} 0 &= f(t)w_0 \\ &= (t^m + a_{m-1}t^{m-1} + \cdots + a_1 t + a_0)w_0 \\ &= tw_{m-1} + a_{m-1}w_{m-1} + \cdots + a_1 w_1 + a_0 w_0, \end{aligned}$$

we see that the action of t on w_{n-1}:

$$tw_{m-1} = -a_{m-1}w_{m-1} - \cdots - a_1 w_1 - a_0 w_0.$$

Therefore, since T acts as left multiplication by t, we have

$$T(w_0) = w_1, T(w_1) = w_2, ..., T(w_{m-2}) = w_{m-1},$$

and

$$T(w_{m-1}) = -a_{m-1}w_{m-1} - \cdots - a_1 w_1 - a_0 w_0.$$

This exactly means that the matrix of T w.r.t the basis $(w_0, w_1, ..., w_{m-1})$ is

$$T(w_0, w_1, ..., w_{m-1}) = (w_0, w_1, ..., w_{m-1})M_f,$$

where M_f is the companion matrix of the polynomial

$$f(t) = t^m + a_{m-1}t^{m-1} + \cdots + a_1 t + a_0 :$$

$$M_f = \begin{bmatrix} 0 & 0 & \cdots & 0 & -a_0 \\ 1 & 0 & \cdots & 0 & -a_1 \\ 0 & 1 & \cdots & 0 & -a_2 \\ \vdots & \vdots & & \vdots & \vdots \\ 0 & 0 & \cdots & 0 & -a_{m-2} \\ 0 & 0 & \cdots & 1 & -a_{m-1} \end{bmatrix}.$$

REMARK 3.22. *We note that the characteristic polynomial* $\det(M_f - tI)$ *of* M_f *is equal to* $f(t)$.

Secondly, we note that each $F[t]$-submodule W_i is a T-invariant F-subspace of the F-vector space V. Thus, if we choose a basis $\mathbf{B}_i$ for each T-invariant subspace W_i and put them together to have a basis $\mathbf{B} = (\mathbf{B}_1, ..., \mathbf{B}_k)$ of V, then the matrix of T looks something like an identity matrix but with block form:

$$\begin{bmatrix} M_1 & & & & \\ & M_2 & & & \\ & & \cdot & & \\ & & & \cdot & \\ & & & & M_k \end{bmatrix},$$

where each M_i is the matrix of the restriction of T to W_i. This proves

THEOREM 3.16. *Let T be an endomorphism on a finite-dimensional vector space V over a field F. Then there exists a basis for V with respect* *to which the matrix of T consists of blocks of the type:*

$$[-a_0], \begin{bmatrix} 0 & -a_0 \\ 1 & -a_1 \end{bmatrix}, \ldots, \begin{bmatrix} 0 & 0 & .. & 0 & -a_0 \\ 1 & 0 & .. & 0 & -a_1 \\ 0 & 1 & .. & 0 & . \\ . & . & .. & . & . \\ . & . & & . & . \\ 0 & 0 & .. & 0 & -a_{m-2} \\ 0 & 0 & .. & 1 & -a_{m-1} \end{bmatrix}.$$

COROLLARY 3.21. *The characteristic polynomial* $\det(T - tI)$ *of T is the product of the polynomials $f_i(t)$, called the* ***invariant factors*** *of T, each of which is associated to the companion matrix M_{f_i}'s.* □

REMARK 3.23. *We note that the invariant factors of T is exactly the same as those d_i's (not in part b but) in part a of Theorem 3.12. Thus, the problem is all about how to compute these invariant factors, and by Theorem 3.15 we know that they are computed from the matrix*

$$M - tI,$$

where M is the matrix of T with respect to some basis of V. Then from Theorem 3.16 and part a of Theorem 3.12 we determine the list *of the invariant factors of the endomorphism T. Moreover, Remark 3.18 says that this list is uniquely determined.*

DEFINITION 3.28. *Such a form in this theorem for the matrix of a linear operator is called a* ***rational canonical form****.*

We note that these blocks are located along the diagonal entries of the identity matrix, and it is the best form available over an arbitrary field, namely, over a field which is not algebraically closed.

The following example tells us how one can obtain such a form from a given matrix for a linear operator T.

EXAMPLE 3.19. *As a continuation of Example 3.7, we have seen Example 3.17. Since*

$$V \approx F[t]/(g(t)) \approx F[t]/(t-1)^2 \times F[t]/(t-2),$$

and since $(t-1)^2 = t^2 - 2t + 1$, *the rational canonical form of the matrix of the linear operator on* V *in Example 3.7 is:*

$$\left[\begin{array}{cc|c} 0 & -1 & \\ 1 & 2 & \\ \hline & & 2 \end{array}\right],$$

where the matrix in the first block is the companion matrix of $(t-1)^2$ *and the matrix in the second block is the companion matrix of* $t-2$.

Here, we note that the invariant factor of V *is* $g(t)$ *by part a of Theorem 3.12.*

EXAMPLE 3.20. *(MAPLE) The following lines compute the rational canonical form of a given matrix:*

```
——————————
interface(echo=2);
with(linalg);
a:=matrix([[0,2,0,-6,2],[1,-2,0,0,2],[1,0,1,-3,2], \
[1,-2,1,-1,2],[1,-4,3,-3,4]]);
reduct1:=ratform(a,'Q');
print(Q);
evalm(Q^(-1)&*a&*Q);
minpoly(a,x);
charpoly(a,x);
simplify(charpoly(a,x)/minpoly(a,x));
——————————
```

6.4. Jordan form. In this subsection, we further specialize our discussion when the bottom field F is the field $\mathbb{C}$ of complex numbers.

Everybody knows that irreducible polynomials in $\mathbb{C}[t]$ are all linear, $p(t) = t - \alpha$ for some $\alpha \in \mathbb{C}$, thus, by part a of Theorem 3.12, every finitely generated $\mathbb{C}[t]$-module V is a direct sum of submodules which are isomorphic to ones of the form

$$\mathbb{C}[t]/(t-\alpha)^n.$$

This is a vector space over $\mathbb{C}$. As before we let w_0 be the residue of 1 in $\mathbb{C}[t]/(t-\alpha)^n$, i.e., $w_0 = 1 + (t-\alpha)^n$.

Thus identifying V, for instance, with $\mathbb{C}[t]/(t-\alpha)^n$, we consider the image of w_0 under the powers $(T-\alpha)^i$ of the linear operator,

$$(T-\alpha) : V \longrightarrow V,$$

$$\begin{aligned} w_1 &= (t-\alpha)w_0, \\ w_2 &= (t-\alpha)^2 w_0, \\ &\cdots \\ w_{n-1} &= (t-\alpha)^{n-1} w_0. \end{aligned}$$

By a similar argument we have done before, we see that the vectors

$$w_0, w_1, ..., w_{n-1}$$

are linearly independent over $\mathbb{C}$, but

$$w_n = (t-\alpha)^n w_0 = 0$$

in $\mathbb{C}[t]/(t-\alpha)^n$. Thus, we have

$$\begin{aligned} Tw_i &= w_{i+1} + \alpha w_i, \qquad i = 0, ..., n-2, \\ Tw_{n-1} &= \alpha w_{n-1}. \end{aligned}$$

This establishes

THEOREM 3.17. *Let T: $V \longrightarrow V$ be an endomorphism on a finite-dimensional complex vector space V. Then there exists a basis of V with* *respect to which the matrix of T consists of blocks of the type:*

$$[\alpha], \begin{bmatrix} \alpha & 0 \\ 1 & \alpha \end{bmatrix}, \begin{bmatrix} \alpha & 0 & \cdots & 0 & 0 \\ 1 & \alpha & \cdots & 0 & 0 \\ 0 & 1 & \cdots & 0 & 0 \\ \vdots & \vdots & & \vdots & \vdots \\ 0 & 0 & \cdots & \alpha & 0 \\ 0 & 0 & \cdots & 1 & \alpha \end{bmatrix}.$$

REMARK 3.24. *Explanation on how to obtain these blocks, or, equivalently, the list of invariant factors, is given in Remark 3.21 and Remark 3.23.*

DEFINITION 3.29. *Such a matrix in this theorem is called a* ***Jordan form****, or a* ***Jordan matrix****.*

Since a Jordan matrix J is lower triangular, we easily see that its eigenvalue is α, and its associated eigenspace

$$E_a = \{v \in V : JX = \alpha X\}$$

is generated by a single eigenvector. In this subsection we thus have established:

COROLLARY 3.22. *Given a complex matrix A there exists an invertible complex matrix P such that PAP^{-1} is a Jordan matrix, which is unique up to permutation of the blocks.* □

EXAMPLE 3.21. *The Jordan form or Jordan matrix of the module in Example 3.17 consists of two Jordan blocks:*

$$\left[\begin{array}{cc|c} 0 & 1 & \\ 1 & 1 & \\ \hline & & 2 \end{array}\right].$$

We note that computing the Jordan form of a given matrix requires finding eigenvalues, i.e., roots of its characteristic polynomial $p(t)$:

$$p(t) = \det(tI - A) = 0.$$

If the eigenvalues are all distinct, the Jordan form is the diagonal matrix:

$$\begin{bmatrix} \alpha_1 & & & & \\ & \cdot & & & \\ & & \cdot & & \\ & & & \cdot & \\ & & & & \alpha_n \end{bmatrix}.$$

Otherwise, let $\alpha_1 = \alpha$ be an r-multiple root of the characteristic equation: $p(t) = 0$. Then depending on $r = 1, 2, 3, \ldots$, there are various possibilities for the part of the Jordan matrix with diagonal entries α.

They can be distinguished by computing eigenvectors of various operators

$$(\alpha I - A)^i$$

related to T.

REMARK 3.25. *This is about **generalized eigenvectors** of an endomorphism A.*

In fact, it can be shown that

PROPOSITION 3.12. *The dimension of the null spaces of the operators*

$$(\alpha I - A)^i, \qquad i = 1, 2, \ldots, r/2,$$

distinguishes the Jordan forms in all cases. □

EXAMPLE 3.22. *When $r = 4$, we have in total 5 different possibilities, basically depending on the number of distinct eigenvectors:*

$$(i) \begin{bmatrix} \alpha & & & \\ 1 & \alpha & & \\ & 1 & \alpha & \\ & & 1 & \alpha \end{bmatrix}, (ii) \left[\begin{array}{ccc|c} \alpha & & & \\ 1 & \alpha & & \\ & 1 & \alpha & \\ \hline & & & \alpha \end{array}\right], (iii) \left[\begin{array}{cc|cc} \alpha & & & \\ 1 & \alpha & & \\ \hline & & \alpha & \\ & & 1 & \alpha \end{array}\right],$$

$$(iv) \left[\begin{array}{cc|c|c} \alpha & & & \\ 1 & \alpha & & \\ \hline & & \alpha & \\ \cline{3-4} & & & \alpha \end{array}\right], (v) \left[\begin{array}{c|c|c|c} \alpha & & & \\ \cline{1-2} & \alpha & & \\ \cline{2-3} & & \alpha & \\ \cline{3-4} & & & \alpha \end{array}\right].$$

Thus, the dimensions of eigenspaces are $1, 2, 2, 3$, *and* 4. *Thus this dimension distinguishes all cases except the second and the third cases. But, these remaining two cases can be distinguished by the matrix* $(\alpha I - A)^2$. *It is zero in the third case and not zero in the second case.*

EXAMPLE 3.23. *(MAPLE) The following lines compute the Jordan form of a given matrix:*

```
————————
interface(echo=2);
with(linalg);
a:=matrix([[1,-1,1],[1,-1,1],[1,-1,0]]);
reduct1:=jordan(a,'P');
print(P);
evalm(P^(-1)&*a&*P);
minpoly(a,x);
charpoly(a,x);
————————
```

Exercises for Section 3.6

EXERCISE 3.24. *Determine the Jordan form of the matrix*

$$A = \begin{bmatrix} 1 & 1 & 0 \\ 0 & 1 & 0 \\ 0 & 1 & 1 \end{bmatrix}.$$

EXERCISE 3.25. *Determine the Jordan form of the matrix*

$$A = \begin{bmatrix} 1 & 1 & 1 \\ -1 & -1 & -1 \\ 1 & 1 & 1 \end{bmatrix}.$$

CHAPTER 4

Field Extensions

1. Introduction

Here is our first example of a field extension.

EXAMPLE 4.1. *We obviously know that the equation $x^2+1=0$ for an unknown x has no real solutions; hence x^2+1 is an irreducible polynomial over $\mathbb{R}$. The following arguments show that this is not a dead end. Rather, it opens a whole new beginning of the so-called Field Theory, provided that we are well equipped with some basic abstract Algebra:*

Since $\mathbb{R}[x]$ is a PID, it is a UFD; hence irreducible elements are prime elements, and vice versa. Thus, by Proposition 2.6, the principal ideal (x^2+1) in $\mathbb{R}[x]$ generated by x^2+1 is a maximal ideal; hence equivalently the quotient ring $\mathbb{R}[x]/(x^2+1)$ is a field.

We then recall that there is a natural ring homomorphism

$$\varphi\colon \mathbb{R}[x] \longrightarrow \mathbb{R}[x]/(x^2+1),$$

which sends $f(x) \longmapsto \bar{f}(x) := f(x) \bmod (x^2+1)$. In particular, the unknown x is sent to $x \bmod (x^2+1) = x$ under this map and the polynomial x^2+1 is sent to 0 under this map. Then since φ is a ring homomorphism, the image $\bar{x}$ of x under φ, which is x, in $\mathbb{R}[x]/(x^2+1)$ is a solution of $(x)^2+1=0$. That is, x in the field $\mathbb{R}[x]/(x^2+1)$ is no longer a relation free object, unlike its preimage, x in $\mathbb{R}[x]$.

On the other hand, the composition of it after the natural injection: $\mathbb{R} \hookrightarrow \mathbb{R}[x]$ gives a natural ring homomorphism

$$\mathbb{R} \hookrightarrow \mathbb{R}[x] \longrightarrow \mathbb{R}[x]/(x^2+1).$$

Then since any ring homomorphism from a field is injective, which is the case because there are only two ideals $(0), (1)$ *in a field, and moreover since, by construction, any real number* r *is sent to itself under this composition, we have constructed a field* $\mathbb{R}[x]/(x^2+1)$ *containing* $\mathbb{R}$ *as its subfield. In other words, we have successfully extended the field* $\mathbb{R}$ *of real numbers to a newly constructed field* $\mathbb{R}[x]/(x^2+1)$*, where the equation* $x^2+1=0$ *indeed has a solution! We customarily write* i *for such an* x *in the field* $\mathbb{R}[x]/(x^2+1)$*: That is,* i *is by definition a number if ever exists whose square is equal to* -1*.*

Our construction explicitly shows that $\bar{x} := x \operatorname{mod}(x^2+1) = x$ *in the field* $\mathbb{R}[x]/(x^2+1)$ *is such an* i*. Then certainly* $-i \in \mathbb{R}[x]/(x^2+1)$ *is another solution of the same equation* $x^2+1=0$ *for an unknown* x*.*

Moreover, $\mathbb{R}[x]/(x^2+1)$ *is obviously a real vector space of dimension* 2*, since* $\{1, i\}$ *is its basis. This is because* $\mathbb{R}[x]$ *is a real vector space of infinite dimension with a basis* $\{1, x, x^2, ...\}$ *and every power* x^j *for* $j > 1$ *is reduced to* ± 1 *or* $\pm x$ *under modulo* (x^2+1)*.*

This allows to write any element of the extended field $\mathbb{R}[x]/(x^2+1)$ *as a* $\mathbb{R}$*-linear combination of* 1 *and* i*,*

$$\mathbb{R}[x]/(x^2+1) = \{a+bi : a, b \in \mathbb{R}\}.$$

In this way, we have constructed the field

$$\mathbb{C} = \{a+bi : a, b \in \mathbb{R}\} = \mathbb{R}[x]/(x^2+1)$$

of complex numbers as a 2*-dimensional real vector space by "algebraically" adjoining an element* i *to* $\mathbb{R}$*.*

Since our construction of the field $\mathbb{C}$ *of complex numbers is canonical, whoever construct it, it must be isomorphic to ours.*

REMARK 4.1. *It is known that Kronecker (1823 - 1891) was the first who constructed the field* $\mathbb{C}$ *of complex numbers by the same procedure described in this example. In Section 5 of this Chapter, we will present essentially the same construction for different fields.*

Definition 4.1. *A field K is **algebraically closed** if every polynomial in one variable over K has a root in it.*

Then Gauss proved that

Theorem 4.1. *(**Fundamental Theorem of Algebra**) The field $\mathbb{C}$ of complex numbers is algebraically closed.* □

We will not provide the proof of this theorem in this book.

Lemma 4.1. *An algebraically closed field is an infinite field.*

Proof. Let k be a finite field consisting of s elements $a_1, ..., a_s$. Then the following polynomial

$$f(x) = 1 + \prod_{i=1}^{s}(x - a_i)$$

has no roots in k at all. Therefore, a finite field cannot be algebraically closed. □

In Chapter one, we briefly discussed how our numbers were extended from the set $\mathbb{N}$ of natural numbers, and mentioned that the main motivation of such extensions was to figure out solutions of polynomial equations in one variable.

We recall that construction of the field $\mathbb{R}$ of real numbers, which is nothing to do with this motivation, took quite long time, although we use them in every approximation related activities. But, then thanks to Gauss (1777 - 1855), if we works within the field $\mathbb{C}$ of complex numbers, we will have no arithmetic difficulties in solving a polynomial equation in one variable. But, since any non-zero element in any field is invertible, i.e., a unit element, if we work only in a field it completely destroys an important role of irreducible elements or prime elements in multiplicative structure of numbers.

Additive structure of numbers is abstracted or generalized to the concept of modules or vector spaces and it was used in many part of mathematics, and this may be one of the reasons why the subject Linear Algebra is very important.

On the other hand, if we do not have enough theory on multiplicative structure which basically stems from that of the set $\mathbb{N}$ of natural numbers, we would not be able to construct the field $\mathbb{C}$ of complex numbers in a very "reasonable" manner, as we have done in our first example above. Because of this reason, we have studied elementary theory of factorizations in the previous chapter spending quite some time. Factorizability is a core concept for examining multiplicative structure of numbers, so the point we must keep in our mind while we study Algebra is, for instance, that,

- **a:** either whenever we have to destroy multiplicative structure in our "extension" business, we try not to destroy too much but only the least needed amounts,
- **b:** or if it turns out that we extend it not too much, then we define a meaningful subring, but is not a subfield, of the extended field by some "algebraic" property, and study its multiplicative structure.

Otherwise, we will suddenly loose too much of algebraic structures we usually have, starting with $\mathbb{N}$. Turning every non-zero numbers into invertible by extending too much makes us arithmetically very easy, but it destroys all other goods things based on multiplicative properties of numbers. We recall Theorem 2.3:

THEOREM 4.2. ***(Fundamental Theorem of Arithmetic)*** *Every non-zero integer a can be written as a product*

$$a = cp_1 \cdots p_k,$$

where $c = \pm 1$, the p_i are prime numbers, and $k \geq 0$. This expression is unique except for the ordering of the prime factors p_i.

Therefore, we are confined in the realm of these two fundamental theorems in "Arithmetic and Algebra" = "Computational" Algebra.

Before we leave this section, we consider a few tiny lemmas which we used in the above example, and will be used in our later discussion whenever needed.

LEMMA 4.2. *In any field F, there are only two ideals,* (0) *and* (1).

PROOF. Let I be a non-zero ideal in a field F. It contains a non-zero element a, and hence by definition of an ideal, $1 = a^{-1}a \in I$. Thus, any ideal in F other than the zero ideal (0) is equal to $(1) = F$. □

We can say more:

PROPOSITION 4.1. *A commutative ring R is a field if and only if it has only two ideals,* (0) *and* (1).

PROOF. Since every non-zero element of a field is invertible, the $\Longrightarrow$ direction holds. Conversely, suppose that R has only two ideals. Since the "zero" ring, $R = \{0\}$, has only one ideal, our ring R under consideration is not the zero ring. Thus the only two ideals are necessarily (0) and (1), because these two are ideals of any ring R. To prove that every non-zero element of R is invertible, let a be a non-zero element of R. Then the principal ideal (a) generated by a is necessarily equal to (1) and hence a multiple of a is 1. That is, a is invertible. □

LEMMA 4.3. *Any ring homomorphism from a field (to a non-trivial ring) is injective.*

PROOF. Let $f : F \longrightarrow R$ be a ring homomorphism and consider its kernel $\ker(f) := f^{-1}(0)$. One easily checks that it is an ideal in F; hence it must be either (0) or $(1) = F$. But, our definition of a ring homomorphism says that $f(1) = 1$, and $1 \neq 0$ in any non-trivial ring R, and hence, $\ker(f) = (0)$, which is equivalent to f being injective. □

LEMMA 4.4. *Any field of characteristic* 0 *is infinite, but not conversely.* □

REMARK 4.2. *Since the algebraic closure of a finite field is an infinite field, the converse of this lemma does not hold in general.*

PROPOSITION 4.2. *There exists exactly one ring homomorphism*

$$\varphi : \mathbb{Z} \longrightarrow R,$$

from the ring $\mathbb{Z}$ of integers to any commutative ring R, which is defined by

$$\begin{aligned} \varphi(n) &= \text{“}n\text{” times } 1_R = 1_R + \cdots + 1_R \ (n \text{ times}) \text{ if } n \geq 1, \\ \varphi(-n) &= -\varphi(n). \end{aligned}$$

PROOF. By definition, any ring homomorphism $\varphi : \mathbb{Z} \longrightarrow R$ sends 1 to 1_R and $n+1$ to $\varphi(n) + \varphi(1)$. Thus, it is (uniquely) determined on the natural numbers by the recursive definition

$$\varphi(1) = 1 \qquad \text{and} \qquad \varphi(n') = \varphi(n) + 1,$$

where n' denotes the successor function: $n' = n + 1$.Thus, this formula, together with $\varphi(-n) := -\varphi(n)$ if $n > 0$ and $\varphi(0) = 0$, determines φ uniquely. Then it is very easy to prove that the map φ is indeed a ring homomorphism. □

Exercises for Section 4.1

EXERCISE 4.1. *Construct a field consisting of only 2 elements.*

EXERCISE 4.2. *Construct a field consisting of only 3 elements.*

EXERCISE 4.3. *Construct a field consisting of only 4 elements.*

2. Examples of Fields

Much of the theory of fields has to do with a pair $F \subset K$ of fields, one contained in the other, where F is considered to be the basic (i.e., given) field, and K is related to it.

DEFINITION 4.2. *An* ***extension field*** *or an* ***extended field*** *of F is a field which contains F as a subfield.*

An extension K of a field F is always a vector space over F. Addition is just the addition law in K, and scalar multiplication of an element α of K by an element c of F is the product $c\alpha$ in K.

We have three important classes of fields.

DEFINITION 4.3.

(1) *a **number field** K is a subfield of $\mathbb{C}$.*
(2) *A **finite field** is a field with finitely many elements.*
(3) *A **function field** is an extended field of the field $\mathbb{C}(X)$ of rational functions.*

Thus, any number field K contains 1 in $\mathbb{C}$, and consequently it contains $\mathbb{Q}$ as a subfield. That is, it is an extension of the field $\mathbb{Q}$ of rational numbers. Among number fields, algebraic number fields are most commonly studied. Among these, real or imaginary quadratic number field

$$\mathbb{Q}[\sqrt{d}],$$

where d is a square free positive or negative integer, is much studied.

For finite fields, we need:

PROPOSITION 4.3. *Every finite field K contains as a subfield $\mathbb{Z}_p = \mathbb{Z}/p\mathbb{Z}$ for some prime integer p.*

PROOF. We know that there is a unique (structural) ring homomorphism $\varphi : \mathbb{Z} \longrightarrow K$. Since $\mathbb{Z}$ is a *PID*, its kernel, $\ker(\varphi) = \varphi^{-1}(0)$, is a principal ideal in$\mathbb{Z}$, so it is generated by a natural number p. Then since K is a field it is automatically an integral domain, and hence the zero ideal (0) in K is a prime ideal. Thus, its inverse image $\ker(\varphi)$ is also a prime ideal in $\mathbb{Z}$. Thus, its generator p must be a prime integer. Then finally by a universally repelling property of the quotient ring $\mathbb{Z}/p\mathbb{Z}$, the homomorphism is decomposed to

$$\mathbb{Z} \longrightarrow \mathbb{Z}/p\mathbb{Z} \longrightarrow K,$$

where the second map is automatically injective. $\square$

DEFINITION 4.4. *The prime field $\mathbb{Z}_p = \mathbb{Z}/p\mathbb{Z}$ is usually denoted by $\mathbb{F}_p$.*

COROLLARY 4.1. *A finite field has a prime characteristic p.* $\square$

Now for function fields.

EXAMPLE 4.2. *A simple example of a function field is the field* $\mathbb{C}(x)$ *of rational functions, which is by definition,*

$$\mathbb{C}(x) \;=\{\textit{equivalence classes of fractions } \frac{f(x)}{g(x)} : f(x), 0 \neq g(x) \in \mathbb{C}[x]\},$$

That is, it is a field of fractions of a UFD, $\mathbb{C}[x] =$ *the set of all complex polynomials in one variable.*

A function field can be defined by an irreducible complex polynomial $f(x, y)$ in two variables, i.e., not a polynomial in just one variable x, say $f(x, y) \in \mathbb{C}[x, y] = \mathbb{C}[x][y]$. For example, let $f(x, y) = y^2 - x^3 + x$. We know that its locus, i.e. its real solution set: $y^2 = x(x-1)(x+1)$, is traditionally called an elliptic curve. Here are graphs of some of them:

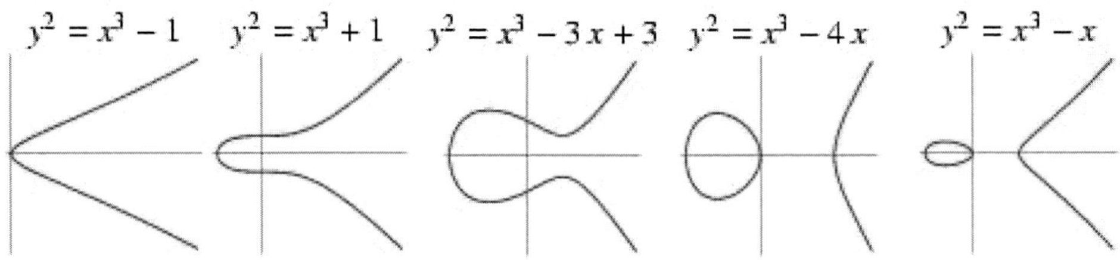

Since it is irreducible in $\mathbb{C}[x][y]$, it is also irreducible in $\mathbb{C}(x)[y]$. Here we note that $\mathbb{C}(x)$ is a field of fractions of a *UFD* $\mathbb{C}[x]$, so we can fully apply Theorem 2.8. Since $\mathbb{C}(x)[y]$ is a *PID*, the ideal $(f(x, y))$ in $\mathbb{C}(x)[y]$ generated by the polynomial $f(x, y)$ in y is a maximal ideal; hence the quotient ring

$$\mathbb{C}(x)[y]/(f(x, y))$$

is a field containing $\mathbb{C}(x)$ as its subfield.

We note that this construction is essentially the same as the one done explicitly presented in our first example in the previous chapter. Of course, the extended field, the functions field $\mathbb{C}(x)[y]/(f(x, y))$ is a finite dimensional $\mathbb{C}(x)$ vector space of dimension $=$ the degree, $\deg(f(x, y))$ of $f(x, y)$ in y. We will study the degree of a field extension very soon.

REMARK 4.3. *We note that all extension fields of finite fields and all function fields are those "constructed" by a process which is very similar to (or spiritually exactly the same as) the procedure explicitly described in our first example in the previous section, and this is the main reason why we started this Chapter with that particular and very important Example, as long as "constructibility" of a mathematical object is concerned or considered to be important.*

Function fields play an important role in the Theory of Analytic Functions, i.e., Complex Geometry, and in Algebraic Geometry, and it may be a good place to mention why such fields are called "function" fields.

Set theoretically, function and maps are defined by the same axiom. But their actual usage in mathematics has not been the same, and will not be the same.

We traditionally use the terminology, maps or mappings, when we want to compare two mathematical objects, i.e., two sets, whereas we traditionally use the terminology, functions, when we want to "evaluate" the numeric "values" in k of points in a space, k^n, where k is usually a field. Any subset S of such a space k^n is usually called a "**geometric object**".

In this latter interpretation, we usually assume that we know very well about the field k, and by "restricting" such functions to a subset of k^n and studying all the fibers, i.e., the inverse images, under the restriction of such functions, of points in k, we hope to have a lot better understanding of the geometric object S.

Among functions definable on k^n, are there polynomials and in particular they are defined purely algebraically, i.e., only using arithmetic. Thus, by naturally identifying the set of variables $\{x_1, ..., x_n\}$ in polynomials in n variables with the set of coordinates functions $\{x_1, ..., x_n\}$ for those points $(a_1, ..., a_n)$ in k^n, we "can" establish a bridge between Algebra and Geometry. Of course, this bridge requires some additional requirements, in order to constitute a "strong" or "solid" bridge.

Thus, considering polynomials in n variables as polynomial "functions" on k^n is conceptually very crucial for both Algebra and Geometry, and moreover, one can easily prove that if k is an infinite field, then polynomial functions are polynomials and conversely polynomials are polynomial functions. That is,

PROPOSITION 4.4. *Let k be an infinite field. Then, two polynomials in n variables with coefficients in k are the same if and and only if they are the same as "functions" defined on k^n.* $\square$

Moreover, polynomial functions are always continuous and differentiable at any point!

Going back to our extended field $\mathbb{C}(x)[y]/(f(x,y))$ of the field $\mathbb{C}(x)$, we see that the residue $\bar{y}$ of the variable y, which is still y, under the ring homomorphism

$$\mathbb{C}(x)[y] \longrightarrow \mathbb{C}(x)[y]/(f(x,y)),$$

does satisfy the original "algebraic" relation:

$$f(x,\bar{y}) = 0,$$

although the original variable y in the polynomial ring $\mathbb{C}(x)[y]$ never ever satisfies such an algebraic relation, where since x is in the bottom field we regard it as a "known" unknown in this equation, and hence "locally" $\bar{y} = y$ is considered as a function $y(x)$ of x, and hence the residue $\bar{y} = y$ and its all other elements of $\mathbb{C}(x)[y]/(f(x,y))$ are "functions" of x This is because $\mathbb{C}(x)[y]/(f(x,y))$ is a vector space over $\mathbb{C}(x)[y]$ with a basis consisting of 1 and finitely many powers y^j of $\bar{y} = y$, as we have seen in our first example in the previous section.

Because of this underlying reason which connect Algebra with Geometry very tightly, they are traditionally called "function" fields, rather than called, say, polynomial fields.

REMARK 4.4. *We know that the Cartesian product $k^n = k \times k \times \cdots k$ (n times) is equipped with a vector space structure, and hence it is called a "space". But since Linear Algebra in general cannot handle non-flat geometric objects, i.e., those other than lines, plane, and hyperplanes etc., in k^n, we need to have a more "stronger" Algebra, i.e., a non-Linear*

Algebra, where multiplication of elements are allowed. Prof. Oscar Zariski was the first who recognized the importance of this point of view while he was at The Johns Hopkins University, and he started a modern Algebraic Geometry at around 1930*s which is based on Commutative Algebra.*

Exercises for Section 4.2

EXERCISE 4.4. *Let F be a field containing exactly* 8 *element. Prove that the characteristic of F is* 2*.*

EXERCISE 4.5. *Why is a field a PID?*

3. Algebraic and Transcendental Elements

DEFINITION 4.5. *Let K be an extension of a field F, and an element α of K is called* ***algebraic*** *over F if it is a root of some non-zero polynomial with coefficients in F. Otherwise, it is called* ***transcendental*** *over F.*

Thus every element α of F is automatically algebraic over itself, because it is a root of $x - \alpha$, which has coefficients in F. Moreover, the unknown x or the variable x in the field $F(x)$ of rational functions = the field of fractions of the *UFD* $F[x]$ is automatically transcendental over F.

On the other hand, we note that the two properties, algebraic and transcendental, are generally dependent on the given bottom field F.

EXAMPLE 4.3. *The complex number $2\pi i$ is algebraic over the field $\mathbb{R}$ of real numbers, but it is transcendental over the field $\mathbb{Q}$ of rational numbers.*

We also note that for a given $\alpha \in K$, the injection $F \hookrightarrow K$ gives rise to the natural ring homomorphism, called the **substitution homomorphism**:

$$\varphi : F[x] \longrightarrow K\,, \tag{4.1}$$

defined by sending $x \mapsto \alpha$, and hence sending $f(x)$ to $f(\alpha)$, and whose restriction on F is the identity map. Thus, in terms of this map, $\alpha \in K$

is transcendental if and only if $\ker(\varphi) = 0$, and is algebraic if and only if $\ker(\varphi) \neq 0$.

If $\alpha \in K$ is algebraic over F, then since $F[x]$ is a *PID*, its kernel $\ker(\varphi)$, which is an ideal in $F[x]$, is generated by a non-zero single element $f(x)$, the monic polynomial of lowest degree having α as a root. Since $f(\alpha) = 0$ in a field K, $f(x)$ must be necessarily an irreducible polynomial over F. Thus, $f(x)$ is the only irreducible monic polynomial in $\ker(\varphi)$, because all other polynomials in $\ker(\varphi)$ are multiples of $f(x)$. We call this polynomial $f(x)$ the **irreducible polynomial for α over F**, or the **minimal polynomial for α over F**.

REMARK 4.5. *It is important to note that this irreducible polynomial $f(x)$ is dependent on F as well as on α, because irreducibility of a polynomial is dependent on the bottom field.*

EXAMPLE 4.4. *Let $F = \mathbb{Q}[i]$, and let α be the complex number $\sqrt{i} = \frac{1}{2}\sqrt{2}(1+i)$. Then the irreducible polynomial for α over $\mathbb{Q}$ is x^4+1, but this polynomial factors in F: $x^4+1 = (x^2+i)(x^2-i)$; hence the irreducible polynomial for α over F is $x^2 - i$.*

Thus, where there are several fields around us, we must make it sure to which field we refer. It is always better to say that $f(x)$ is **irreducible over F**, or that it is an **irreducible element of $F[x]$**.

DEFINITION 4.6. *Let $\alpha \in K$. The field extension of F which is generated by an element $\alpha \in K$ is denoted by*

$$F(\alpha).$$

Thus by definition, $F(\alpha)$ is the smallest field containing F and α. This is true because it is the intersection of all subfields of K containing both K and α.

More generally, if $\alpha_1, ..., \alpha_n$ are elements of an extension field K of F, then the notation

$$F(\alpha_1, ..., \alpha_n)$$

will denote for the smallest subfield of K containing both F and the elements $\alpha_1, ..., \alpha_n$.

We denote the ring generated by α over F by

$$F[\alpha].$$

Thus, it consists of those elements of K which can be written as a polynomial in α with coefficients in F:

$$a_n\alpha^n + a_{n-1}\alpha^{n-1} + \cdots + a_1\alpha + a_0,$$

with $a_i \in F$. Then since $F[\alpha]$ is automatically an integral domain, we have its field FR of fractions, and moreover, the subfield $F(\alpha)$ is isomorphic to the field FR of fractions of $F[\alpha]$ under the map

$$FR \ni a/b \longmapsto ab^{-1} \in F(\alpha),$$

where $a, b \in F[\alpha]$. Thus, elements of $F(\alpha)$ are ratios of elements of the form (3).

Now if $\alpha \in K$ is transcendental over F, then $\varphi(F[x]) = F[\alpha]$ is isomorphic to $F[x]$, where $\varphi : F[x] \longrightarrow K$ is the map in (4.1).

PROPOSITION 4.5. *If α is transcendental over F, then the map*

$$F[x] \longrightarrow F[\alpha]$$

in (Cf. 4.1) is an isomorphism, and hence $F(\alpha)$ is isomorphic to the field $F(x)$ of rational functions. □

This simple fact has the consequence that the field extensions $F(\alpha)$ are isomorphic for all transcendental element α, because they are all isomorphic to the field of rational functions $F(x)$.

EXAMPLE 4.5. *We recall that π and e are both transcendental over $\mathbb{Q}$. Therefore, $\mathbb{Q}(\pi)$and $\mathbb{Q}(e)$ are isomorphic fields, sending π to e.*

REMARK 4.6. *We know how these two transcendental numbers are defined. But, both of them are transcendental simply means that there is no way of differentiating them algebraically.*

PROPOSITION 4.6.

a: *Suppose that α is algebraic over F, and let $f(x)$ be its irreducible polynomial over F. Then the map*

$$F[x]/(f(x)) \longrightarrow F[\alpha]$$

is an isomorphism, and $F[\alpha]$ is a field. Thus, $F[\alpha] = F(\alpha)$.

b: *More generally, let $\alpha_1, ..., \alpha_n$ be algebraic element of a field extension K of F. Then $F[\alpha_1, ..., \alpha_n] = F(\alpha_1, ..., \alpha_n)$.*

PROOF. Let φ be the map in (4.1), with $K = F(\alpha)$. Since $f(x)$ generates $\ker(\varphi)$, $F[x]/(f(x))$ is isomorphic to its image, which is $F[\alpha]$. Since $f(x)$ is irreducible over F, it generates a maximal ideal, and hence $F[\alpha]$ is in fact a field; hence in particular, its field of fractions is itself. Then since $F(\alpha)$ is the smallest subfield of K containing both F and α, and $F[\alpha]$ is already a subfield of K, $F(\alpha) = F[\alpha]$. This proves part a. For part b, we note that $F[\alpha_1, ..., \alpha_n] = F[\alpha_1, ..., \alpha_{n-1}][\alpha_n]$ and hence induction on n will finish its proof. □

PROPOSITION 4.7. *Let α be an algebraic element over F, and let $f(x)$ be its irreducible polynomial over F. Suppose $f(x)$ has degree n. Then $(1, \alpha, \alpha^2, ..., \alpha^{n-1})$ is a basis for $F[\alpha]$ as a vector space over F.*

PROOF. Since we have a projection $F[x] \longrightarrow F[x]/(f(x))$, which sends $g(x)$ to $g(x) \bmod (f(x))$, and since $F[x]$ is a vector space over F with a basis $(1, x, x^2, ...)$ the images $\bar{x}^j = x^j \bmod f(x) = x^j$ of the x^j under this projection for $j = 0, 1, ..., n-1$ is a basis of $F[x]/(f(x))$. For this, one just checks the images of x^j are (i) linearly independent over F and (ii) they span $F[x]/(f(x))$. Then since the isomorphism $F[x]/(f(x)) \longrightarrow F[\alpha]$ is induced from the ring homomorphism $F[x] \longrightarrow F[\alpha]$, which sends x to α and is identity on F, $(1, \alpha, ..., \alpha^{n-1})$ is a basis of the F-vector space $F[\alpha]$. For this, we have to use the relation $f(\alpha) = 0$ in $F[\alpha]$. □

Now, let α, β be two algebraic elements over F. If $F[\alpha]$ is isomorphic to $F[\beta]$, then the irreducible polynomial for α over F and the irreducible polynomial for β necessarily have the same degree over F, because this degree is the dimension of the field extension as an F-vector space. But, as the following example shows, having the same degree is not sufficient for different reasons.

EXAMPLE 4.6. *Let* $f(x) = x^2 + 1$, *and* $g(x) = x^2 + 5$. *We have already seen that in the field* $\mathbb{Q}[x]/(f(x))$ *is the ring* $\mathbb{Z}[i]$ *of Gauss integers, which is the collection of all algebraic integers over* $\mathbb{Z}$ *there, has 4 units:* $\pm 1, \pm i$.*But, in the field* $\mathbb{Q}[x]/(g(x)) = \mathbb{Q}[\sqrt{-5}]$ *is the ring* $\mathbb{Z}[\sqrt{-5}]$, *which is the collection of all algebraic integers over* $\mathbb{Z}$ *there, has only* 2 *units:* ± 1.*Hence, these two fields can not be isomorphic. In fact,* $\mathbb{Z}[i]$ *is a PID but* $\mathbb{Z}[\sqrt{-5}]$ *is not a UFD, so it is not a PID.*

On the other hand, one easily checks that if α *is a root of* $x^3 - x + 1$, *then* $\beta = \alpha^2$ *is a root of* $x^3 - 2x^2 + x - 1$. *[Hint:* $\alpha^3 - \alpha = -1$ *and square it.] Since their reduction* $\bmod 2$ *is the same irreducible polynomial* $x^3 + x + 1$ *over the field* $\mathbb{Z}_p = \mathbb{Z}/p\mathbb{Z}$ *with* $p = 2$, *these two polynomials are irreducible over* $\mathbb{Z}$, *and hence irreducible over* $\mathbb{Q}$, *by Corollary 2.5. We obviously have* $\mathbb{Q}[\beta] \subseteq \mathbb{Q}[\alpha]$.*Then since* $\mathbb{Q}[\alpha]$ *is a* $\mathbb{Q}$*-vector space of dimension* 3 *and* $\mathbb{Q}[\beta]$ *is its subspace with the same dimension* 3, *two fields* $\mathbb{Q}[\alpha]$ *and* $\mathbb{Q}[\beta]$ *are actually the same.*

REMARK 4.7. *This example could confuse us. But, if we examine very carefully ring homomorphisms we have used so far, we can recognize that, when restricted, they are the identity map on the bottom field* F, *by our same method of construction of extended fields of* F *in terms of irreducible polynomials over* F, *as is in the first example of this Chapter.*

DEFINITION 4.7. *Let* K, L *be two extended fields of a field* F. *A ring homomorphism (isomorphism, respectively)* $\varphi : K \longrightarrow L$ *is called an* F-***homomorphism*** *or an* F*-algebra homomorphism(an* F***-isomorphism*** *or an* F*-algebra isomorphism, respectively) if its restriction on the bottom field* F *is the identity map. The two extended fields* K, L *of the same field* F *are called* ***isomorphic field extensions*** *if there is an* F*-isomorphism between them.*

PROPOSITION 4.8. *Let $\varphi : K \longrightarrow L$ be an F-isomorphism of field extensions of a field F, and let $f(x)$ be a polynomial with coefficients in F. Let α be a root of $f(x)$ in K, and let $\beta = \varphi(\alpha)$ be its image in L. Then β is also a root of the same polynomial $f(x)$.*

PROOF. Let $f(x) = a_n x^n + \cdots + a_1 x + a_0$. Then $\varphi(a_i) = a_i$ and $\varphi(\alpha) = \beta$. Then since φ is a ring homomorphism which is the identity map on F, we can expand as follows:

$$\begin{aligned} 0 &= \varphi(0) = \varphi(f(\alpha)) \\ &= \varphi(a_n\alpha^n + \cdots + a_1\alpha + a_0) \\ &= \varphi(a_n)\varphi(\alpha)^n + \cdots + \varphi(a_1)\varphi(\alpha) + \varphi(a_0) \\ &= a_n\beta^n + \cdots + a_1\beta + a_0. \end{aligned}$$

□

If we bring these facts into our discussion, we have the following important result. In our later sections we are going to use this result quite often.

PROPOSITION 4.9. *Let K, L be two extended fields of the same field F. Let $\alpha \in K$ and $\beta \in L$ be* algebraic *elements. Then the following statements are equivalent.*

a: *There exists an isomorphism of fields*

$$\sigma : F(\alpha) \xrightarrow{\sim} F(\beta)$$

which is identity map on the bottom field F and which sends α to β.

b: *The irreducible polynomials for α and β over F are the same.*

PROOF. The $\Longrightarrow$ direction holds by the above proposition. The converse holds because we have two F-isomorphisms:

$$F[x]/(f(x)) \xrightarrow{\sim} F[\alpha] \qquad \text{and} \qquad F[x]/(f(x)) \xrightarrow{\sim} F[\beta].$$

□

EXAMPLE 4.7. *We know, for example, by Eisenstein Criterion (Theorem 2.11 of Chapter 2) that the polynomial* $x^3 - 2$ *is irreducible over* $\mathbb{Q}$. *Let* α *be its real root, and let* $\zeta = e^{2\pi i/3}$ *be a complex cubic root of* 1. *Then, since*

$$(\frac{x}{\alpha})^3 = 1,$$

$\alpha, \zeta\alpha$, *and* $\zeta^2\alpha$ *are the three complex roots of* $x^3 - 2$. *Thus, by the above proposition, we have a* $\mathbb{Q}$*-isomorphism*

$$\mathbb{Q}(\alpha) \xrightarrow{\sim} \mathbb{Q}(\zeta\alpha),$$

sending α *to* $\zeta\alpha$. *In this case,* $\mathbb{Q}(\alpha)$ *is a subfield of* $\mathbb{R}$, *but* $\mathbb{Q}(\zeta\alpha)$ *is not. Thus, in order to properly understand this isomorphism, we must stop viewing these fields as subfields of* $\mathbb{C}$ *and we only look at their internal algebraic structure.*

Exercises for Section 4.3

EXERCISE 4.6. *Determine the irreducible polynomial for* $\alpha = \sqrt{3}+\sqrt{5}$ *over each of the following fields.*

(1) $\mathbb{Q}$.
(2) $\mathbb{Q}(\sqrt{3})$.
(3) $\mathbb{Q}(\sqrt{10})$.
(4) $\mathbb{Q}(\sqrt{15})$.

EXERCISE 4.7. *Let* $f(x) = x^3 - 3x + 4 \in \mathbb{Q}[x]$.

(1) *Prove that* $f(x)$ *is irreducible over* $\mathbb{Q}$.
(2) *Determine the inverse of* $\alpha^2+\alpha+1$ *in* $\mathbb{Q}(\alpha)$ *in the form* $a+b\alpha+c\alpha^2$, *with* $a, b, c \in \mathbb{Q}$.

4. The Degree of a Field Extension

An extension K of a field F is always a vector space over F. Addition is just the addition law in K, and scalar multiplication of an element α of K by an element c of F is the product $c\alpha$ in K.

DEFINITION 4.8. *The dimension of K as an F-vector space over F is called the* **degree** *of the field extension $F \subset K$. It is denoted by*

$$[K : F],$$

and an extension $F \subset K$ is called a **finite extension** *if $[K : F]$ is finite.*

The degree is the simplest but important invariant of an extension of a field.

EXAMPLE 4.8. *$\mathbb{C}$ has the $\mathbb{R}$-basis $(1, i)$, and hence $[\mathbb{C} : \mathbb{R}] = 2$.*

Extensions of degree 2 are called *quadratic extensions,* those of degree 3 are called *cubic*, and so on.

We note that the degree of an extension $F \subset K$ is 1 if and only if $F = K$.

In fact, the term degree comes from the case that $K = F(\alpha)$ is generated by one algebraic element over F. In this case, K has a basis

$$(1, \alpha, \alpha^2, ..., \alpha^{n-1}),$$

where n is the degree of the irreducible polynomial for α over F by Proposition 4.7. This establishes the first important property of the degree.

PROPOSITION 4.10. *If α is algebraic over F, then $[F(\alpha) : F]$ is the degree of the irreducible polynomial for α over F.* □

This degree is also called the **degree of α over F**. Thus, we have:

COROLLARY 4.2. *α has degree 1 if and only if it is an element of the bottom field F, and α has degree ∞ if and only if it is transcendental over F.*

PROOF. The proof of the first part is very easy. For the second part, if α is transcendental over F then the subfield $F[\alpha]$ of $F(\alpha)$ is infinite dimensional and hence so is $F(\alpha)$. If α is algebraic over F, then by Proposition 4.7, α has a finite degree. □

Now, the smallest non-trivial degree is 2, which is also a prime integer. In this case, we can completely describe all the quadratic field extension of a field F.

PROPOSITION 4.11. *Suppose that a field F does not have characteristic 2, that is, $1+1 \neq 0$. Then any extension $F \subset K$ of degree 2 is obtained by adjoining a square root: $K = F(\delta)$, where $\delta^2 = D$ is an element of F. Conversely, if δ is an element of an extension field of F, such that $\delta^2 \in F$ but $\delta \notin F$, then $F(\delta)$ is a quadratic extension.*

PROOF. The last assertion of the proposition is obvious.

We first show that every quadratic extension is obtained by adjoining a root of a quadratic polynomial $f(x) \in F[x]$. For this, let α be any element of K but it is not in F. Then $(1, \alpha)$ is clearly linearly independent over F. Since K has dimension 2 as an F-vector space, $(1, \alpha)$ is a basis of K over F, and $K = F[\alpha]$. It follows that α^2 is a linear combination of $(1, \alpha)$, say $\alpha^2 = -b\alpha - c$, with $b, c \in F$. Then α is a root of $f(x) = x^2 + bx + c$.

Now, we change $F[\alpha]$ to the "standard" form $F[\delta]$ using the quadratic formula. Here, since $2 \neq 0$ in F, we can use the quadratic formula

$$\alpha = \frac{1}{2}(-b + \sqrt{b^2 - 4c})$$

to solve the equation $x^2+bx+c = 0$. This is checked by direct calculation. There are two choices for the square root, and one of which gives our chosen root α. Define $\delta = \sqrt{b^2 - 4c}$. Thus, $\delta \notin F$ but $\delta^2 \in F$. We need to show that $\delta \in K$: If we write β for another root of $x^2 + bx + c = 0$, Then $x^2 + bx + c = (x - \alpha)(x - \beta)$, and from this equality we see that $\delta = \alpha - \beta = \alpha + (\alpha + b) = 2\alpha + b \in F[\alpha] = K$, and hence it also generates K over F. Its square is the discriminant $b^2 - 4c$, which is in F. Thus, $K = F[\alpha] = F(\delta)$, with $\delta^2 \in F$. □

REMARK 4.8. *Several examples of finite extensions of low degrees are treated in the later chapters, Chapter 5 and Chapter 6.*

When the degree > 2, we have the second important property of the degree: It is multiplicative in towers of fields.

THEOREM 4.3. *Let $F \subset K \subset L$ be fields. Then $[L : F] = [L : K][K : F]$.*

PROOF. Let $\mathbf{B} = (y_1, ..., y_n)$ be a basis for L as a K-vector space, and let $\mathbf{C} = (x_1, ..., x_m)$ be a basis for K as an F-vector space. Thus, $[L : K] = n$ and $[K : F] = m$. We will prove that their mn products $(..., x_i y_j, ...)$ is a basis $\mathbf{P}$ of L as an F-vector space. The same reasoning will work if $\mathbf{B}$ or $\mathbf{C}$ is infinite.

Let α be an element of L. Since $\mathbf{B}$ is a basis for L over K, we can write

$$\alpha = \beta_1 y_1 + \cdots + \beta_n y_n,$$

with $\beta_i \in K$ in a unique way. Then since $\mathbf{C}$ is a basis for K over F, we can uniquely express each β_i as a linear combination of the x_j's:

$$\beta_j = a_{1j} x_1 + \cdots + a_{mj} x_m,$$

with unique $a_{ij} \in F$. Thus,

$$\alpha = \sum_{i,j} \sum a_{ij} x_i y_j,$$

shows that the products $x_i y_j$ spans L as an F-vector space. Given α, we have seen that each β_i is uniquely determined and each a_{ij} is again uniquely determined from each β_j; hence the a_{ij} are uniquely determined for a given α. Thus, $\mathbf{P} = (..., x_i y_j, ...)$ is a basis for L over F. □

One important case of a tower of field extensions is that K is a given extension of a field F and α is an element of K. Then $F(\alpha)$ generated by α is an *intermediate* field:

$$F \subset F(\alpha) \subset K.$$

Here, α could be an element of the bottom field F. Now, if we add one more finite condition that $[K : F]$ is finite to theses settings, we have an immediate corollary:

COROLLARY 4.3. *Let α be an element of a finite extension K of F of degree n. Then α is algebraic over F, and its degree divides n.*

PROOF. Since the whole degree $[K : F]$ is finite, its divisor $[F(\alpha) : F]$ must be also finite by Theorem 4.3; hence α is algebraic over F of degree $[F(\alpha) : F]$ by the second part of Corollary 4.2. □

DEFINITION 4.9. *An extension field K of a field F is called an* ***algebraic extension*** *if every element of K is algebraic over F, and K is said to be* ***algebraic over*** *F if it is an algebraic extension of F.*

Thus, this corollary also says:

COROLLARY 4.4. *A finite extension of F is an algebraic extension.*

But, the converse does not hold in general: The algebraic closure $\bar{F}$ of a finite field F is not a finite extension by Lemma 4.1.

THEOREM 4.4. *Let $F \subset K \subset L$ be fields. If L is algebraic over K and K is algebraic over F, then L is algebraic over F.*

PROOF. We have to show that every element $\alpha \in L$ is algebraic over F. Since α is algebraic over K, it satisfies some equation of the form

$$\alpha^n + a_{n-1}\alpha^{n-1} + \cdots a_1\alpha + a_0 = 0,$$

with coefficients $a_0, ..., a_{n-1} \in K$. Thus, α is algebraic over the subfield $F(a_0, ..., a_{n-1})$ generated by $a_0, ..., a_{n-1}$ over F. Then since each coefficients a_i is algebraic over F, we can consider a tower or an increasing chain of subfields

$$F \subset F(a_0) \subset F(a_0, a_1) \subset \cdots \subset F(a_0, ...a_{n-1}) \subset F(a_0, ...a_{n-1}, \alpha),$$

obtained by adjoining these coefficients a_i, and α in succession. Then since each a_{i+1} is algebraic over F, it is algebraic over $F(a_0, ...a_i)$ and also α is algebraic over $F(a_0, ...a_{n-1})$. Thus, each intermediate extension in the chain is a finite extension. Therefore, by Theorem 4.3, the degree of $F(a_0, ...a_{n-1}, \alpha)$ over F is finite. Thus, by Corollary 4.3 α is indeed algebraic over F. □

Now, we would like to examine some explicit applications of Corollary 4.3.

COROLLARY 4.5. *Let α be an element of a field extension K of F of prime degree p and $\alpha \notin F$. Then α has degree p over F; hence $K = F(\alpha)$.*

PROOF. This is because the prime integer p divides the degree products $[K : F(\alpha)][F(\alpha) : F]$ and $[F(\alpha) : F] > 1$ since $\alpha \notin F$. Thus, $p = [F(\alpha) : F]$. Then since the subfield $F(\alpha)$ has the same dimension as K, it must be the same as K. □

If we further specialize to the case where $F = \mathbb{R}$ and $p = 2$, we obtain

COROLLARY 4.6. *Every irreducible real polynomial has degree* 1 *or* 2.

PROOF. Let $g(x)$ be an irreducible polynomial over $\mathbb{R}$. Then $g(x)$ has a root α in $\mathbb{C}$ by Gauss. Then since $[\mathbb{C} : \mathbb{R}] = 2$, the degree of the intermediate field $\mathbb{R}(\alpha)$ must be either 1 or 2. □

But as the following example shows that the degree related questions could be quite complicated when the bottom field F is smaller than $\mathbb{R}$, i.e., for instance, $F = \mathbb{Q}$.

EXAMPLE 4.9. *Let $\alpha = \sqrt[3]{2}, \beta = \sqrt[4]{5}$. Consider the field $\mathbb{Q}(\alpha, \beta)$, obtained by adjoining α and β to the field $\mathbb{Q}$ of rational number. We are going to explicitly show that $\mathbb{Q}(\alpha, \beta)$ is a finite extension of degree* $12 = 3 \cdot 4$.

Since the irreducible polynomials for α, β over $\mathbb{Q}$ respectively are $x^2 - 2$ and $x^4 - 5$, $[\mathbb{Q}(\alpha) : \mathbb{Q}] = 3$ and $[\mathbb{Q}(\beta) : \mathbb{Q}] = 4$. Thus, $\mathbb{Q}(\alpha, \beta)$ contains subfields of degree 3 *and* 4*, respectively; hence $[\mathbb{Q}(\alpha, \beta) : \mathbb{Q}]$ must be a common multiple of* 3 *and* 4*. Then since $x^4 - 5$ may not be irreducible over $\mathbb{Q}(\alpha)$, $[\mathbb{Q}(\alpha, \beta) : \mathbb{Q}(\alpha)]$ is at most* 4*. Thus,*

$$[\mathbb{Q}(\alpha, \beta) : \mathbb{Q}(\alpha)][\mathbb{Q}(\alpha) : \mathbb{Q}] = [\mathbb{Q}(\alpha, \beta) : \mathbb{Q}] \leq 12.$$

But since no number < 12 is both divisible by 3 *and* 4*, $[\mathbb{Q}(\alpha, \beta) : \mathbb{Q}]$ must be* 12.

REMARK 4.9. *We note that* $\gcd(3, 4) = 1$ *in this example; hence,*

$$\text{lcm}(3, 4) = 3 \cdot 4 / \gcd(3, 4) = 12.$$

Thus, the arguments made in this example should work when $\gcd(n, m) = 1$ *in general.*

Here is a lot more complicated example:

EXAMPLE 4.10. *Let* $f(x) = x^4 + 2x^3 + 6x^2 + x + 9$. *Then since*

$$f(x) \bmod 2 = x^4 + x + 1$$

is irreducible over $\mathbb{Z}_2$, *it is irreducible over* $\mathbb{Q}$. *[Cf. Example 2.16]. Let* $\alpha = \sqrt[3]{2}$ *and* γ *be a root of* $f(x)$.

REMARK 4.10. *If one actually solves* $f(x)$ *for* x, *the* 4 *solutions are all non-real complex numbers and they are not real multiples of* $i = \sqrt{-1}$.

Then we easily see that $\alpha \notin \mathbb{Q}(\gamma)$ *since* $3 = [\mathbb{Q}(\alpha) : \mathbb{Q}]$ *never divides* $4 = [\mathbb{Q}(\gamma) : \mathbb{Q}]$. *Thus,* $\mathbb{Q}(\alpha)$ *is never a subfield of* $\mathbb{Q}(\gamma)$.

On the other hand, it is not easy to check whether i *is in* $\mathbb{Q}(\gamma)$, *although* i *has degree* 2 *over* $\mathbb{Q}$.

CLAIM 4.1. $i \notin \mathbb{Q}(\gamma)$.

One way of verifying this claim is the following:

Suppose $i \in \mathbb{Q}(\gamma)$. *Then since* $\mathbb{Q}(\gamma)$ *is a vector space with a basis* $\{1, \gamma, \gamma^2, \gamma^3\}$, $i = a + b\gamma + c\gamma^2 + d\gamma^3$ *for uniquely determined rational numbers* $a, b, c, d \in \mathbb{Q}$. *Then*

$$-1 = i^2 = (a + b\gamma + c\gamma^2 + d\gamma^3)^2 = A + B\gamma + C\gamma^2 + D\gamma^3,$$

where A, B, C *and* D *are obviously quadratic polynomials in* a, b, c, d, *and all* γ^j *for* $j > 3$ *are reduced to lower degree terms using* $\gamma^4 + 2\gamma^3 + 6\gamma^2 + \gamma + 9 = 0$.

Then since $\{1, \gamma, \gamma^2, \gamma^3\}$ *is a* $\mathbb{Q}$*-basis of* $\mathbb{Q}(\gamma)$, $(A+1) + B\gamma + C\gamma^2 + D\gamma^3 = 0$ *implies* $a, b, c,$ *and* d *must satisfy a system of polynomial equations for* a, b, c, d:

$$(*) \begin{cases} A + 1 &= 0 \\ B &= 0 \\ C &= 0 \\ D &= 0 \end{cases}.$$

That is, this system must have rational solutions if $i \in \mathbb{Q}(\gamma)$.

Therefore, if we can show that this system does not have any rational solutions, then it just means that $i \notin \mathbb{Q}(\gamma)$.

From actual symbolic computations (using the Computer Algebra System REDUCE), the above procedure gives us the following system of polynomial equation for a, b, c, d:

$$(*) \begin{cases} a^2 - 18bd - 9c^2 + 36cd - 36d^2 + 1 & = 0 \quad (1) \\ 2ab - 2bd - c^2 - 14cd + 14d^2 & = 0 \quad (2) \\ 2ac + b^2 - 12bd - 6c^2 + 22cd - 31d^2 & = 0 \quad (3) \\ 2ad + 2bc - 4bd - 2c^2 - 4cd + 3d^2 & = 0 \quad (4) \end{cases}.$$

If $d = 0$, *then* $2 \times (2) - (4)$ *gives* $4ab = 2bc$ (5). *Now if* $d = 0$ *and* $b = 0$, *then* (4) *gives:* $c = 0$; *hence* $a = \pm i$. *If* $d = 0$ *but* $b \neq 0$, *then* (5) *gives:* $2a = c$ *so that* (1) *gives:* $35a^2 = 1$, *but there is no such* a *in* $\mathbb{Q}$. *Thus, if* $d = 0$, *we cannot have rational solutions for* $(*)$. *What then can you do when* $d \neq 0$?

If one solves this system of quadratic equations in a, b, c, d, *using* ***the Theory of Groebner basis***, *one can easily check that it has no rational solutions. Therefore,* $i \notin \mathbb{Q}(\gamma)$; *hence* $\mathbb{Q}(i)$ *is not a subfield of* $\mathbb{Q}(\gamma)$.

In fact, if we examine the following solution set carefully, we recognize that d *is a solution of an irreducible polynomial over* $\mathbb{Q}$; *hence* d *is not a rational number; hence so are others.*

```
answer := {
{d=root_of(222751263710994018575*d_**14
+ 32724804681752629920*d_**12
+ 1024333138422735392*d_**10 - 6646593532140800*d_**8
+ 100778855020288*d_**6 + 742952112128*d_**4
- 1618280448*d_**2 - 16777216,d_,tag_4),
c=(d*( - 83748769052826334863019725\
115741370601846002617736357021762\
346872149077225*d**12-11290968900\
828018089493846300962145523331020\
```

```
460782200955244042491473112560*d**10
- 24357197840754268165241460520265\
431864678998131101052507725339652\
2745056*d**8 + 6082720643139816979\
3939255499088507623359245008701692\
64936927101866496*d**6
- 103487770448543458846949576351951\
299957358332200257397456304798091520*d**4
+ 305377420121353933353783686617421\
653272989311661892922663630770176*d**2
+325157226594771408443395112261694117\
6991191389079565891588784128))/10687223\
42559005143419558568990116082618392523\
378061991375011840,
b=(d*(339674522264481233280307226498433\
736958200024095604577412257987889496382175*d**12
+ 5022445084682341196453410288092040895\
1778395326730667791206173045622940480*d**10
+ 16092858531302623135393969520717052710950\
00887324050674074332022108532448*d**8
-86663404289530880657901092591433959135\
69181418487710710553564357026048*d**6
+ 14293837772695056785045942236867738\
3959217706297401849329111670813440*d**4
+ 130092449352402613432539374341841864\
0892318676859644856444540915712*d**2
+42954530524522914922748671167150203563\
12065440986576180077592576))/10687223425590051\
43419558568990116082618392523378061991375011840,
a=(d*(2340757785015665201575523972012705078\
4998336919351934069102976168784876675*d**12
+ 3499707692073841937688591239838394352274\
```

```
272379279511880610227383192782680*d**10
+ 11517333894855290895628017175310501 1\
3217957243960397979630883274794551 68*d**8
-618556881002408988524647090616746493824 21\
719497972036262937237938 7008*d**6
+ 30458426905531619358808852805900543596 2\
787561210667259734018274944 0*d**4
+ 15586036997419414144587973317287135426\
32898501099852334991579013 12*d**2
-55201348856723044104108201208662058997041617\
352501989897528934 4))/9293237761382653421039 63\
973034883550102950020328749557717401 60},
{d=0,c=0,b=0,a=i},
{d=0,c=0,b=0,a= - i}
}$
```

End of Example. □

REMARK 4.11. *This example shows us usefulness of exact computation, i.e., symbolic computation in studying Abstract Algebra. At the end of this Section we are going to have more examples which should convince the reader the important role of the Theory of Groebner basis.*

The previous theorem and the next theorem state the most important *abstract* consequences of the multiplicative property of degrees.

THEOREM 4.5. *The collection of all algebraic elements over F in an extension field K of a field F is a subfield of K.*

PROOF. Let α, β be two algebraic elements of K. We must show that $\alpha + \beta, \alpha\beta, -\alpha$, and α^{-1} (if $\alpha^{-1} \neq 0$) are all algebraic too. Since α is algebraic over F, $[F(\alpha) : F] < \infty$. Then since β is also algebraic over F, β is also algebraic over the bigger field $F(\alpha)$. Thus, $F(\alpha, \beta)$ is generated by β over $F(\alpha)$ so that it is a finite extension of $F(\alpha)$, i.e., $[F(\alpha, \beta) : F(\alpha)] < \infty$. Thus, by Theorem 4.3 $[F(\alpha, \beta) : F]$ is finite too.

Then since every finite extension is an algebraic extension, we see that $F(\alpha, \beta)$ is algebraic over F. Then since $\alpha+\beta, \alpha\beta, -\alpha$, and α^{-1} (if $\alpha \neq 0$) are in $F(\alpha, \beta)$, they are all algebraic. □

REMARK 4.12. *Now suppose that the irreducible polynomials for α and β over F are given to us. Then by this proof, we know that $\alpha+\beta$ is algebraic over F. But, the proof does not tell us how to compute the irreducible polynomial for $\alpha+\beta$ over F.*

We are going to discuss a computational method, which enables us to compute a polynomial over F whose root is $\alpha+\beta$. This method is historically known as **a method of undetermined coefficients**.

Suppose that the irreducible polynomials for α of degree d_1 and β of degree d_2 over field F are known. Let $n = d_1 d_2$. Then Since $F(\alpha, \beta) = F[\alpha, \beta]$, each element of $F(\alpha, \beta)$ is an F-linear combination of the n monomials $\alpha^i \beta^j$, $0 \leq i \leq d_1 - 1, 0 \leq j \leq d_2 - 1$. Thus, given $\gamma \in F(\alpha, \beta)$, we write the powers $1, \gamma, \gamma^2, ..., \gamma^n$ as F-linear combinations of these monomials. Since we have $n+1$ number of the powers γ^j and only n monomials $\alpha^i \beta^j$, the powers must be linearly dependent; hence a linear dependence relation among the powers γ^j determines a polynomial with coefficients in F of which root is γ. Thus, we only need to figure out a linear dependence relation among the powers γ^j by relating these powers in one way or another.

Here is such an example.

EXAMPLE 4.11. *Suppose that $\alpha = \sqrt{a}, \beta = \sqrt{b}$, where $a, b \in F$. We want to determine a polynomial whose root is $\gamma = \alpha + \beta$. To do this, we square γ and square its square, etc., and try to relate these powers together in order to find a linear relations among the powers γ^j.*

$$\begin{aligned} \gamma^2 &= (\alpha+\beta)^2 = (a+b) + 2\alpha\beta, \\ (\gamma^2)^2 &= (a^2 + 6ab + b^2) + 4(a+b)\alpha\beta. \end{aligned}$$

In this case, these two equations are enough since we can relate these two equations by getting rid of $\alpha\beta$*:*

$$\gamma^4 - 2(a+b)\gamma^2 + (a-b)^2 = 0.$$

Thus, γ *is a root of the polynomial*

$$g(x) = x^4 - 2(a+b)x^2 + (a-b)^2$$

over F.

But there is one point which complicates matters. Let $g(x)$ be the polynomial having γ as a root which we find in this way. This polynomial may be reducible. For instance, it may happen that γ is actually in the bottom field F, though α, β are not in F. If so, the method of undermined coefficients described above is unlikely to produce its irreducible equation $x - \gamma$. It is hard to determine the *irreducible* polynomial for γ over F.

REMARK 4.13. *In the above it says that it is hard to determine the irreducible polynomial for* $\gamma = \alpha + \beta$ *for a given algebraic elements of* K *over* F*. Yes, it is hard if we do not have a strong computational means.*

The thing is that even though we study field extensions in terms of polynomials in one variable over a field F*, we some times have to explicitly figure out certain coefficients by solving a system of polynomial equations, which we have seen in Example 4.10 and in the explanation of the method of unknown coefficients given above.*

We can explicitly compute the irreducible polynomial for γ*, using the theory of Groebner bases, a fairly recently developed computation theory for polynomials in several variables since* 1965*. One of the things that this book aims at is to try to combine effectively axiomatic treatments, whose main purpose is to pursuit certain properties qualitatively, and computational treatments, whose main purpose is to compute and compare everything quantitatively, into a one piece of Algebra. This is because no one can actually apply purely abstract theory if it is only fancy for those axiomists, unless it is well-supported by a theory of actual computation.*

Here are some further examples as applications of the Theory of Groebner bases.

EXAMPLE 4.12. *Let* $\beta = \frac{1+\sqrt{2}}{\sqrt{2}} \in \mathbb{Q}(\sqrt{2})$. *We want to compute the minimal polynomial of* β *over* $\mathbb{Q}$. *For this, we set:*

$$\begin{aligned} f(x) &= x^2 - 2, \\ p(x) &= 1 + x, \\ q(x) &= x. \end{aligned}$$

Then we use the Computer Algebra System, REDUCE:

```
—————————
off echo$
off nat$
load_package groebner$
torder({x,y},lex);
f:=x^2-2;
p:=1+x;
q:=x;
gb:=groebner({f,q*y-p});
gb := {x - 2*y + 2,
2*y**2 - 4*y + 1}$
—————————
```

Thus, $y^2 - 2y + 1/2$ *is the minimal polynomial of* β *over* $\mathbb{Q}$. *We easily check that*

$$\left(\frac{1+\sqrt{2}}{\sqrt{2}}\right)^2 - 2\frac{1+\sqrt{2}}{\sqrt{2}} + \frac{1}{2} = 0.$$

Thus, $\mathbb{Q}(\beta) = \mathbb{Q}(\sqrt{2})$.

REMARK 4.14. *Applying the method of undetermined coefficients, we can get a polynomial* $4x^4 - 12x^2 + 1$ *of which solution is* β. *But it may not be easy to see manually that*

$$4x^4 - 12x^2 + 1 = (2x^2 - 4x + 1)(2x^2 + 4x + 1).$$

REMARK 4.15. *In the above example, we happen to know that x^2-2 is the minimal polynomial of $\sqrt{2}$ over $\mathbb{Q}$. What will happen if we only know that $\sqrt{2}$ is a root of a polynomial over $\mathbb{Q}$? For example, $\sqrt{2}$ is certainly a root of $f=(x^2-x^3)(x^2-2)$. If we compute a Groebner basis for the ideal $(f, q*y-p)$ then do you think we are still able to figure out the minimal polynomial of $\beta=\frac{1+\sqrt{2}}{\sqrt{2}}$ over $\mathbb{Q}$? The answer is Yes:*

EXAMPLE 4.13. *This is a continuation of the above example. Here we replace the previous $f=x^2-2$ with a polynomial multiple*

$$f=(x^2-x+3)(x^2-2)$$

of f in order to see whether we can still figure out the minimal polynomial of $\beta=\frac{1+\sqrt{2}}{\sqrt{2}}$ over $\mathbb{Q}$. We use the Computer Algebra System, REDUCE:

```
off echo$
off nat$
load_package groebner$
torder({x,y},lex)$
f:=(x^2-x+3)*(x^2-2);
p:=1+x;
q:=x;
gb:=groebner({f,q*y-p});
gb := {x - 6*y**3 + 20*y**2 - 21*y + 6,
6*y**4 - 26*y**3 + 41*y**2 - 27*y + 5}$
result:=factorize(part(gb,length(gb)));
result := {{3*y**2 - 7*y + 5,1},{2*y**2 - 4*y + 1,1}}$
on rounded$
result1:=part(sub({y=(1+sqrt(2))/sqrt(2)},part(result,1)),1);
result1:=1.79289321881$
result2=part(sub({y=(1+sqrt(2))/sqrt(2)},part(result,2)),1);
result2 := 0$
```

In the above computation, we actually verify that the last polynomial

$$6y^4 - 26y^3 + 41y^2 - 27y + 5$$

of the computed result, gb, is a multiple of the minimal polynomial of $\beta = \frac{1+\sqrt{2}}{\sqrt{2}}$ *over* $\mathbb{Q}$*, by first factorizing it and then numerically evaluating each of its two factors at* $\beta = \frac{1+\sqrt{2}}{\sqrt{2}}$*. Thus, we conclude that*

$$2y^2 - 4y + 1$$

is indeed the minimal polynomial of $\beta = \frac{1+\sqrt{2}}{\sqrt{2}}$ *over* $\mathbb{Q}$*.*

EXAMPLE 4.14. *Let* $\beta = \frac{1+2\sqrt[3]{3}+3(\sqrt[3]{3})^2}{1+\sqrt[3]{3}}$*. We want to compute the minimal polynomial for* β*. For this we set:*

$$\begin{aligned} f(x) &= x^3 - 3, \\ p(x) &= 1 + 2x + 3x^2, \\ q(x) &= 1 + x. \end{aligned}$$

Then we use the Computer Algebra System, REDUCE:

```
off echo$
off nat$
load_package groebner$
torder({x,y},lex);
f:=x^3-3;
p:=1+2*x+3*x^2;
q:=1+x;
gb:=groebner({f,q*y-p});
gb := {61*x + 2*y**2 - 23*y - 27,
2*y**3 + 3*y**2 - 21*y - 107}$
```

Thus, the minimal polynomial of β *over* $\mathbb{Q}$ *is*

$$y^3 + (3/2)y^2 - (21/2)y - 107/2\,.$$

Thus, $\mathbb{Q}(\beta) = \mathbb{Q}(\sqrt[3]{3})$.

EXAMPLE 4.15. *We will see in Example 5.3 that any* $\gamma = i + c\sqrt[3]{2}$ *with a non-zero* $c \in \mathbb{Q}$ *is a primitive element for* $K = \mathbb{Q}(i, \sqrt[3]{2})$. *Then, for instance, what is the minimal polynomial over* $\mathbb{Q}$ *of the primitive element* $i + \sqrt[3]{2}$*?*

The minimal polynomial of $\sqrt[3]{2}$ *over* $\mathbb{Q}$ *is clearly* $x^3 - 2$*, and since* $\mathbb{Q}(\sqrt[3]{2})$ *is a subfield of* $\mathbb{R}$*, the minimal polynomial* $y^2 + 1$ *of* $i = \sqrt{-1}$ *over* $\mathbb{R}$ *is again the minimal polynomial of* i *over* $\mathbb{Q}(\sqrt[3]{2})$*.*

Then the following codes for the Computer Algebra System, REDUCE, will provide the right answer very quickly:

———————

```
off echo$
off nat$
load_package groebner$
q1:=x^3-2$
q2:=y^2+1$
q3:=z-(x+y)$
result1:=groebner({q1,q2,q3},{x,y,z});
result1:={22*x - 12*z**5 + 9*z**4 - 40*z**3 + 78*z**2 - 100*z - 91,
22*y + 12*z**5 - 9*z**4 + 40*z**3 - 78*z**2 + 78*z + 91,
z**6 + 3*z**4 - 4*z**3 + 3*z**2 + 12*z + 5}$
```

———————

Thus, the last equation in result1 with only variable z*, so changing* z *into* x *for our convience, we see that*

$$x^6 + 3x^4 - 4x^3 + 3x^2 + 12x + 5,$$

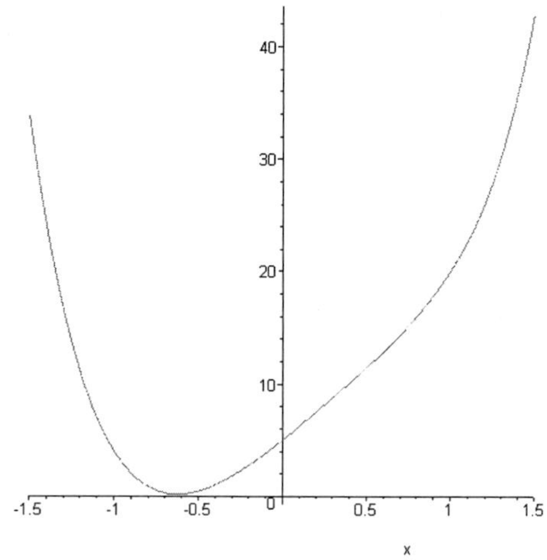

is the minimal polynomial over $\mathbb{Q}$ *of* $i+\sqrt[3]{2}$*, i.e., the irreducible polynomial over* $\mathbb{Q}$ *of least degree of* $i+\sqrt[3]{2}$.

EXAMPLE 4.16. *If we examine the computed reduced Groebner basis of Example above, we immediately see that* $i+\sqrt[3]{2}$ *is a primitive element for* $\mathbb{Q}(i, \sqrt[3]{2})$ *over* $\mathbb{Q}$*, and*

$$\begin{aligned}\sqrt[3]{2} &= \frac{-1}{22}\left(\begin{array}{c} -12\left(i+\sqrt[3]{2}\right)^5+9\left(i+\sqrt[3]{2}\right)^4-40\left(i+\sqrt[3]{2}\right)^3 \\ +78\left(i+\sqrt[3]{2}\right)^2-100(i+\sqrt[3]{2})-91 \end{array}\right), \\ i &= \frac{-1}{22}\left(\begin{array}{c} 12\left(i+\sqrt[3]{2}\right)^5-9\left(i+\sqrt[3]{2}\right)^4+40\left(i+\sqrt[3]{2}\right)^3 \\ -78\left(i+\sqrt[3]{2}\right)^2+78\left(i+\sqrt[3]{2}\right)+91 \end{array}\right).\end{aligned}$$

Carefully examine the polynomial expression on the right side of the two equalities for $\sqrt[3]{2}$ *and* i*, respectively, and then examine the following codes, in order to recognize what we have computed as the reduced Groebner basis in Example 5.5.*

Here are the actual codes for such computation for the REDUCE.

```
off echo$
off nat$
f := ((-1)/22)*( - 12*z**5 + 9*z**4 - 40*z**3 + 78*z**2 - 100*z -
91);
g:=((-1)/22)*(12*z**5 - 9*z**4 + 40*z**3 - 78*z**2 + 78*z + 91);
```

```
result1:= sub({z=i+2^(1/3)},f);
2**(1/3)$
result2:= sub({z=i+2^(1/3)},g);
i$
```

——————————

REMARK 4.16. *We are going to have more examples like these in Chapter 5, and we remark that these results are not obtained accidentally. In our next book,* Algebraic Analysis I, *the theory of Groebner basis behind these examples are well explained in detail.*

Exercises for Section 4.4

EXERCISE 4.8. *Let α be an element of a field $\mathbb{C}$ which generates a field extension F of $\mathbb{Q}$ of degree 5. Prove that α^2 also generates F.*

EXERCISE 4.9. *Write $\omega_n = e^{2\pi i/n}$. Prove that $\omega_5 \notin \mathbb{Q}(\omega_7)$.*

EXERCISE 4.10. *Prove that an extension K of F of degree 1 is equal to F.*

EXERCISE 4.11. *Decide whether or not $i = \sqrt{-1}$ is an element of the following fields.*

(1) $\mathbb{Q}(\sqrt{-2})$.
(2) $\mathbb{Q}((\sqrt[4]{-2})$.
(3) $\mathbb{Q}(\alpha)$, *where* $\alpha^3 + \alpha + 1 = 0$.

EXERCISE 4.12. *Let K be a field generated over F by two elements α, β of relatively prime degrees m, n, respectively. Prove that $[K : F] = mn$.*

5. Symbolic Adjunction of Roots

Except the first example Example 4.1 of this Chapter, we have used subfields of the field $\mathbb{C}$ of complex numbers, so that we did not need to construct them in an abstract manner. We simply adjoin complex numbers to the field $\mathbb{Q}$ of rational numbers as desired and work with the subfield they generate.

Since the abstract construction presented as the first example of this Chapter can be repeated in a general setting, we are going to present in this section such an abstract construction of a brand new field. This is needed to construct all finite fields and function fields, because they are not subfields of the field $\mathbb{C}$ of complex numbers, for instance, each for a different reason.

Let $f(x)$ be a polynomial over a field F. Then there exists a natural projection

$$F[x] \longrightarrow F[x]/(f(x)),$$

defined by sending each polynomial $g(x)$ in $F[x]$ to $g(x) \bmod (f(x))$. In particular, x is mapped to itself under this ring homomorphism and each element of the field F is also mapped to itself under this homomorphism. But the polynomial $f(x)$ is sent to 0, and this exactly means that the residue in $F[x]/(f(x))$ of the "symbol" x in $F[x]$ is a root of the equation $f(x) = 0$ in this canonically constructed quotient ring $K := F[x]/(f(x))$, so that x is "no longer" a symbol at all because it now satisfies an algebraic relation $f(x) = 0$! Since any ring homomorphism from a field is always injective, if we combine the above homomorphism with the trivial injection

$$F \hookrightarrow F[x],$$

which sends each element of F to itself, we see that the composition

$$F \hookrightarrow F[x] \longrightarrow F[x]/(f(x))$$

is also injective and sends each element of F to itself. In this way, we have naturally constructed a ring whose subring is the field F, where $f(x) = 0$ has a root.

On the other hand, the problem of finding a root of a given polynomial $f(x)$ is always reduced to a "smaller" factor of $f(x)$, because handling a smaller polynomial is computationally easy to everybody. This requires factorization theory we have studied in Chapter 2 in detail. According to what we have learned, we know that $F[x]$ is a *PID*, and hence it is a *UFD*. Thus, a polynomial $f(x)$ over a field F is irreducible if and only if the ideal $(f(x))$ generated by $f(x)$ is a maximal ideal in $F[x]$ if and only if $F[x]/(f(x))$ is a field.

We have in fact proved a famous work of Leopold Kronecker (1823-1891):

THEOREM 4.6. *(Kronecker) Let F be a field, and let $f(x)$ be an irreducible polynomial in $F[x]$. Then the ring $K := F[x]/(f(x))$ is an extension of F, and the residue $\bar{x}$ of x is a root of $f(x)$ in K.* □

PROPOSITION 4.12. *Let $f(x)$ be a monic polynomial of positive degree over a field F. Then there exists a field extension K of F such that $f(x)$ factors into linear factors over K.*

PROOF. We use induction on the degree of $f(x)$. As a first case, if $f(x)$ has a root α in the bottom field F, then $f(x) = (x - \alpha)g(x)$ with some polynomial $g(x)$ of less degree than that of $f(x)$. Thus, we replace $f(x)$ with $g(x)$ and we are done by induction hypothesis. Otherwise, we choose an irreducible factor $g(x)$ of $f(x)$. Then by Theorem 4.6 there exists a field extension F_1 of F, in which $g(x)$ has a root α. We replace F with F_1 and we are reduced to the first case. □

REMARK 4.17. *We have seen in the above that the polynomial ring $F[x]$ plays an important role in the study of field extension. Then since we are working with two fields at the same time, there is an interplay between their polynomial rings.*

We note that if K is an extension field of F, then the polynomial ring $F[x]$ is a subring of $K[x]$. Thus, computations made in $F[x]$ are always valid in $K[x]$.

PROPOSITION 4.13. *Let $f(x)$ and $g(x)$ be polynomials with coefficients in a field F and let K be an extension field of F. Then*

- **a:** *Division with remainder of $g(x)$ by $f(x)$ gives the same answer, whether carried out in $F[x]$ or $K[x]$.*

b: *$f(x)$ divides $g(x)$ in $K[x]$ if and only if $f(x)$ divides $g(x)$ in $F[x]$.*

c: *The monic greatest common divisor d of $f(x)$ and $g(x)$ is the same, whether it is computed in $F[x]$ or $K[x]$.*

d: *If $f(x)$ and $g(x)$ have a common root in K, then they are not relatively prime in $F[x]$. Conversely, if $f(x)$ and $g(x)$ are not relatively prime in $F[x]$, then there exists an extension field L in which they have a common root.*

e: *If $f(x)$ is irreducible in $F[x]$ and if $f(x)$ and $g(x)$ have a common root in K, then $f(x)$ divides $g(x)$ in $F[x]$.*

PROOF. Part *a*: We carry out the division in $F[x]$ to get $g(x) = f(x)q(x) + r(x)$. This equation also holds in $K[x]$. Now, suppose that we carry out the division in $K[x]$. Then since coefficients of both $f(x)$ and $g(x)$ are in F, the coefficient of $q(x)$ at each division step and hence that of $r(x)$ at the same step are also in F. This is because F is closed under both multiplication and addition. Thus, $q(x)$ and $r(x)$ are in fact in $F[x]$.

Part *b* is a special case of part *a* with the zero remainder.

For part *c*, let $d(x), d'(x)$ be the monic greatest common divisors of $f(x)$ and $g(x)$ in $F[x]$ and in $K[x]$, respectively. Since $d(x)$ is also a common divisor in $K[x]$, $d(x)$ divides $d'(x)$ in $K[x]$. In addition to this, we know that $d(x)$ has the form $d(x) = p(x)f(x) + q(x)g(x)$, for some elements $p(x), q(x) \in F[x]$. This is because $F[x]$ is a *PID*. Thus, $d'(x)$ divides $d(x)$ in $K[x]$. This means that they are associates in $K[x]$, and, being monic, they are equal.

Part *d*: Let α be a common root of $f(x)$ and $g(x)$ in K. Then $x - \alpha$ is a common divisor of $f(x)$ and $g(x)$ in $K[x]$; hence their greatest common divisor in $K[x]$ is not 1. Thus, by part *c* they are also not relatively prime in $F[x]$ either. Conversely, if $f(x)$ and $g(x)$ are not relatively prime, i.e., they have a common divisor $d(x)$ of degree > 0 in $F[x]$, then by Proposition 4.12, $d(x)$ has a root in some extension field L of F and this root will be a common root of both $f(x)$ and $g(x)$ in $L[x]$.

Part e: If $f(x)$ is irreducible over F, then its only divisors in $F[x]$ are 1, $f(x)$, and their associates. Thus by part d the greatest common divisor $d(x)$ of $f(x)$ and $g(x)$ in $F[x]$ is not 1. Thus, $d(x)$ must be $f(x)$ itself. □

Before we move into the next section, Finite Fields, we want to study an easy means which allows us to handle multiple roots effectively. In analysis, the derivative of a function in one variable is defined as the limit of the so-called Newton quotient of the function. But, when the functions are polynomials in one variable, since they are all continuous we do not need limiting process in defining the derivative of a polynomial function. We simply define

DEFINITION 4.10. *Let $f(x) = a_n x^n + \cdots + a_1 x + a_0$ be a polynomial with coefficients in a field F. The (formal)* ***derivative*** *of $f(x)$ is the linear combination of the (formal) derivatives $(x^n)' = nx^{n-1}$:*

$$f'(x) = a_n n x^{n-1} + a_{n-1}(n-1)x^{(n-1)-1} + \cdots + a_1 \cdot 1 \cdot x^{1-1}.$$

The integer coefficients in this formula are to be interpreted as element of F via the canonical ring homomorphism $\mathbb{Z} \longrightarrow F$ in Lemma 4.3. Thus, the derivative of the $f(x)$ is another polynomial with coefficients in the same field F. Then one can easily show that rules such as the product rule for differentiation hold. In fact, you ask high school students what is the derivative of a polynomial function and how it is derived, it is quite likely that they will repeat the above definition, and no more than that.

For us, the most important property of the derivative for polynomials in one variable is that it can be used to recognize multiple roots of a polynomial in one variable.

LEMMA 4.5. *Let $f(x)$ be a polynomial with coefficients in a field F and α be its root in some extended field K of F. Then α is a multiple root, meaning that $(x-\alpha)^2$ divides $f(x)$, if and only if it is also a root of its derivative $f'(x)$.*

PROOF. Write $f(x) = (x-\alpha)g(x)$ in $K[x]$. Then it is clear that α is a root of $g(x)$ if and only if it is a multiple root of $f(x)$. Then by the product role for differentiation, we have

$$f'(x) = (x - \alpha)g'(x) + 1 \cdot g(x).$$

Therefore $g(\alpha) = 0$ if and only if $f'(\alpha) = 0$. □

PROPOSITION 4.14. *Let $f(x) \in F[x]$ be a polynomial. There exists a field extension K of F in which $f(x)$ has a multiple root if and only if $f(x)$ and $f'(x)$ are not relatively prime in $F[x]$.*

PROOF. The $\Longrightarrow$ direction holds by Lemma 4.5 and the part d of Proposition 4.13. Conversely, if $f(x)$ and $f'(x)$ are not relatively prime, they have a common root in some field extension K of F by the part d of Proposition 4.13 and hence $f(x)$ has a multiple root there by Lemma 4.5. □

Here is the most important application of the derivative to field theory:

PROPOSITION 4.15. *Let $f(x)$ be an* irreducible *polynomial with coefficients in a field F. Then $f(x)$ has no multiple root in any field extension of F unless the derivative $f'(x)$ is the zero polynomial. In particular, if F is a field of characteristic zero, the irreducible $f(x)$ has no multiple root.*

PROOF. We are going to prove this based on Proposition 4.14. For this, suppose that $f'(x)$ is not the zero polynomial. Then since $f(x)$ is irreducible, the only way that it can have a non-constant factor in common with another polynomial $g(x)$ is for $f(x)$ to divide $g(x)$ by part e of Proposition 4.13. Moreover if $f(x)$ does divide $g(x)$ then $\deg(g(x)) \geq \deg(f(x))$, or else $g(x) = 0$. But since the degree of $f'(x)$ is always strictly less than that of $f(x)$, $f(x)$ and its derivative $f'(x)$ cannot have a non-constant common factor in common unless $f'(x) = 0$, as required.

In the field of characteristic zero, the derivative of a non-constant polynomial is never zero. □

REMARK 4.18. *The derivative of a non-constant polynomial $f(x)$ may be identically zero if F has a prime characteristic p. This happens when the exponent of every monomial occurring in $f(x)$ is divisible by prime p.*

EXAMPLE 4.17. *In a field F of characteristic 5, the derivative of*

$$f(x) = x^{25} + ax^{15} + bx^5 + c,$$

where a, b, c are arbitrary element of F, must be always the zero polynomial. Since the derivative of this polynomial is identically zero, its roots in any extended field are all multiple roots, by Lemma 4.5. In this example, whether $f(x)$ is irreducible or not is dependent on both the field F and coefficients a, b, c.

Exercises for Section 4.5

EXERCISE 4.13. *For which field F and which primes p, does the polynomial $x^p - x$ have a multiple root?*

6. Finite Fields

In this section, we are going to establish a theorem on finite fields. As we have already proved it in Proposition 4.3, each finite field K contains one of the prime fields $\mathbb{F}_p = \mathbb{Z}_p = \mathbb{Z}/p\mathbb{Z}$. Since K is a vector space over some prime field $\mathbb{F}_p$, it is a finite-dimensional over $\mathbb{F}_p$; hence it is isomorphic to the $\mathbb{F}_p$-vector space $\mathbb{F}_p^r$, where r is the dimension of K over $\mathbb{F}_p$. Therefore, the order, i.e. the number of elements, of K is p^r, a power of prime p. It is customary to write it as

$$q = p^r = |K|.$$

When we refer to finite fields, p will always denote a prime integer and q a power of p, the number of elements, *order*, of the field.

Like $\mathbb{F}_p$ which is a field with p elements, a field with q elements are often denoted by $\mathbb{F}_q$. We are going to show that all fields with the same number of elements are isomorphic, so this notation is not too ambiguous. However, the isomorphism will not be unique when $r > 1$.

6.1. Examples of Finite Fields.

EXAMPLE 4.18. *A field of two elements. Since the number 2 is prime, $\mathbb{Z}/2\mathbb{Z}$ is such a field.*

EXAMPLE 4.19. *A field of three elements. Since the number* 3 *is prime,* $\mathbb{Z}/3\mathbb{Z}$ *is such a field.*

EXAMPLE 4.20. *A field of* 4 *elements. Since* $4 = 2^2$ *is not prime,* $\mathbb{Z}/4\mathbb{Z}$ *is not even an integral domain because* 2 *is a zero divisor there; hence it is not a field. But, over the field* $\mathbb{F}_2 = \mathbb{Z}/2\mathbb{Z}$*, we have the only irreducible polynomial* $f(x)$ *of degree* 2*:*

$$f(x) = x^2 + x + 1 \in \mathbb{F}_2[x],$$

and hence by "adjoining" a root α *of* $f(x)$ *to* $\mathbb{F}_2$*, we see that*

$$K := F[x]/(x^2 + x + 1)$$

is a field of order 4 *since it is a* 2*-dimensional vector space over* $\mathbb{F}_2$*. Since* $\{1, \alpha\}$ *is its* $\mathbb{F}_2$*-basis,* K *consists of the following* 4 *elements:*

$$\{0, 1, \alpha, 1 + \alpha\}.$$

Of course, $1 + \alpha$ *is another root of* $f(x)$ *in* $K = \mathbb{F}_4$*.*

EXAMPLE 4.21. *In general,* $\mathbb{Z}/p\mathbb{Z}$ *is a field if and only if* p *is a prime integer.*

REMARK 4.19. *On the other hand, the introduction of this section also says that it is impossible to have a field with* pq *elements where* p *and* q *are distinct prime numbers.*

EXAMPLE 4.22. *For distinct prime numbers* p *and* q*,* $\mathbb{Z}/p\mathbb{Z} \times \mathbb{Z}/q\mathbb{Z}$ *with the usual component-wise addition and multiplication cannot be a field, because* $(q, 0)$ *and* $(0, p)$ *are non-zero elements but their product* $(q, 0) \cdot (0, p) = (0, 0)$*, so it cannot be an integral domain.*

As these examples show, we need a construction of finite field of order p^r for $r > 1$ in terms of an irreducible polynomial over the prime field $\mathbb{F}_p$.

6.2. Preliminaries. To present a straightforward proof of the theorem, Theorem 4.7 below, we need a few preliminaries.

LEMMA 4.6. *Let k be a natural number dividing r, say $r = ks$, and let $q = p^r$, $q' = p^k$. Then $x^{q'} - x$ divides $x^q - x$.*

PROOF. Obviously we know that

$$y^d - 1 = (y-1)(y^{d-1} + \cdots + y + 1).$$

Substituting $y = q'$ and $d = s$, we see that $q' - 1$ divides $q - 1 = q'^s - 1$. Thus, with $d' = \frac{q-1}{q'-1}$ we have

$$\begin{aligned} x^q - x &= x(x^{q-1} - 1) \\ &= x((x^{p^k-1})^{\frac{q-1}{q'-1}} - 1) \\ &= x(x^{p^k-1} - 1)((x^{p^k-1})^{d'-1} + \cdots + x^{p^k-1} + 1) \\ &= (x^{q'} - x)((x^{p^k-1})^{d'-1} + \cdots + x^{p^k-1} + 1). \end{aligned}$$

□

PROPOSITION 4.16. *Let L be a field of characteristic p, and let $q = p^r$. Then in the polynomial ring $L[x, y]$, we have*

$$(x+y)^q = x^q + y^q \in L[x,y].$$

PROOF. We first prove it when $q = p$. We recall the Binomial Theorem in $\mathbb{Z}[x, y]$:

$$(x+y)^p = x^p + \binom{p}{1} x^{p-1} y + \cdots + \binom{p}{p-1} x y^{p-1} + y^p \in \mathbb{Z}[x,y]$$

Then since the integers $\binom{p}{j}$ for $0 < j < p$ are all divisible by p (see the proof of Corollary 2.7). Thus its image under the natural map

$$\varphi : \mathbb{Z}[x,y] \longrightarrow L[x,y]$$

is simply

$$x^p + y^p = \varphi((x+y)^p) = (\varphi(x) + \varphi(y))^p = (x+y)^p.$$

Now for a general $q = p^r$ with $r > 1$ we apply induction on r: Suppose that the proposition is proved for numbers less than $r > 1$. Write $q' =$

p^{r-1}. Then by induction hypothesis we have

$$\begin{aligned}(x+y)^q &= ((x+y)^{q'})^p \\ &= (x^{q'}+y^{q'})^p = (x^{q'})^p + (y^{q'})^p \\ &= x^q + y^q.\end{aligned}$$

□

PROPOSITION 4.17. *Let p be a prime, and let $q = p^r$.*

- **a:** The polynomial $x^q - x$ has no multiple root in any field L of characteristic p.
- **b:** Let L be a field of characteristic p containing all the roots of $x^q - x$ and let K be the set of all the roots of $x^q - x$ in L. Then K is a subfield of L.

PROOF. Part a: The derivative of $x^q - x$ is $qx^{q-1} - 1 = -1$, because the coefficient $q = 0$; hence $x^q - x$ and its derivative cannot have any common root; hence by Proposition 4.14, $x^q - x$ has no multiple root.

Part b: The existence of such an L is guaranteed by Proposition 4.12. Let α, β be roots of $x^q - x$. We have to show that $\alpha \pm \beta$, $\alpha\beta$, α^{-1} (if $\alpha \neq 0$) are roots of the same polynomial $x^q - x$. It is very clear that $\alpha\beta$ and α^{-1} are also roots of $x^q - x$; and since $(\alpha \pm \beta)^q = \alpha^q \pm \beta^q$ by the above Proposition 4.16. □

In the field $\mathbb{C}$ of complex numbers, the n-th roots of unity clearly form a cyclic group of order n, generated by

$$\zeta = e^{2\pi i/n}$$

In general, it is true that the n-th roots of unity in any field F form a cyclic group. But, its proof is not very easy:

PROPOSITION 4.18. *Let F be (a finite or an infinite) field, and let H be a finite subgroup of the multiplicative group $F^\times := F - \{0\}$, of order n. Then H is a cyclic group, and it consists of all the n-th roots of unity in F.*

PROOF. If H has order n, then the order of *any* element α of H always divides n; hence α is an n-th root of unity, a root of the polynomial $x^n - 1$ in $F[x]$. Then since this polynomial has at most n roots in F (see Proposition 2.1), there aren't any other roots in F, other than those elements of H. Thus, H consists of all the n-th roots of unity in F.

It is harder to prove that H is cyclic: The Structure Theorem for abelian groups (Theorem 3.9 or Corollary 3.12), which states that H is isomorphic to a direct product of cyclic groups:

$$H = \mathbb{Z}/d_1\mathbb{Z} \times \cdots \times \mathbb{Z}/d_k\mathbb{Z},$$

where $d_1|d_2 \cdots |d_k$ and $n = d_1 \cdots d_k$, where $d_1|d_2$ means d_1 divides d_2, for instance. Thus, the order of any element of this product divides d_k; hence every element of H is a root of

$$x^{d_k} - 1.$$

But we already know that this polynomial has at most d_k roots in F, and since H consists of $n = d_1 d_2 \cdots d_k$ elements, the only possibility is that $n = d_k$, $k = 1$, and H is cyclic. □

In particular, this proposition says that the multiplicative group $K^\times$ of order $q-1$ itself is a cyclic group, consisting of all the $(q-1)$-th roots of unity.

REMARK 4.20. *This proposition does not tell us how to figure out a cyclic generator of H.*

6.3. A theorem. Here are the main facts about finite fields, which is due to Galois (1811 - 1832).

THEOREM 4.7. *(Galois) Let p be a prime, ad let $q = p^r$ with $r \geq 1$. Then*

- **a:** *There exists a field of order $q = p^r$.*
- **b:** *Let K be a field of q elements, i.e., of order q. (Remark: Existence of such a field is in part a). Then the elements of K*

are roots of the polynomial $x^q - x$*. This polynomial has distinct roots, and it factors into linear factors in* K*.*

c: *Let* K *be a field* K *of order* q*. Then the multiplicative group* $K^\times$ *of non-zero elements of* K *is a cyclic group of order* $q-1$*.*

d: *Any two fields of order* q *are isomorphic.*

e: *Every irreducible polynomial of degree* r *in* $\mathbb{F}_p[x]$ *is a factor of* $x^q - x$*, and the irreducible factors of* $x^q - x$ *in* $\mathbb{F}_p[x]$ *are precisely the irreducible polynomials in* $\mathbb{F}_p[x]$ *whose degree divides* r*.*

f: *A field* K *of order* q *contains a subfield of order* $q' = p^k$ *if and only if* k *divides* r*.*

PROOF. Part *a*: Let $f(x)$ be $x^q - x$. Then by Proposition 4.12 there exists a field L, containing $\mathbb{F}_p$ as a subfield; moreover the polynomial $x^q - x$ factors into linear factors in L. Then the roots in L of this polynomial form a subfield K of order q, of L by part *b* of Proposition 4.17.

Part *b*: Let K be a field of order q. Its multiplicative group $K^\times = K - \{0\}$ has order $q-1$, i.e., the number of elements of $K^\times$ is $q-1$. Then the order of *any* element α in $K^\times$, i.e., the number of elements of the cyclic subgroup generated by α, divides $q-1$ by Lagrange's Theorem in Group Theory. Thus we have $\alpha^{q-1} = 1$. This means that α is a root of the polynomial $x^{q-1} - 1$. Thus, every element of $K = K^\times \cup \{0\}$ is a root of $x(x^{q-1} - 1) = x^q - x$. Then since $x^q - x$ and its derivative $qx^{q-1} - 1 = -1$ (see the proof of Proposition 4.17) they cannot have a common divisor; hence the roots of $x^q - x$ are all distinct in K, so it factors into linear factors in K:

$$x^q - x = \prod_{\alpha \in K} (x - \alpha) \in K[x],$$

and this proves part *b*.

Part *c*: By an *n-th root of unity* in a field F, we mean an element α whose n-th power is 1. Thus, α is an n-th root of 1 if and only if it is a root of a polynomial

$$x^n - 1,$$

or if and only if its order m, as an element of the multiplicative group $K^\times$, divides n. Here, the $\Leftarrow$ direction of the last equivalence is obvious, and if we write $n = am+r$ with $0 \le r < m$ for unique quotient a and remainder $0 \le r < m$ by the division algorithm then, since m is the smallest number such that $\alpha^m = 1$, we must have $r = 0$, and this proves the $\Rightarrow$ direction.

Then by part b of the theorem, we already know that elements of $K^\times$ are all distinct, and hence if there *exists* a root α of $x^{q-1} - 1$ of order $q-1$, then the powers α^j for $j = 1, ..., q-1$ must be all distinct; hence the subgroup generated by this α must be the same as $K^\times$; consequently $K^\times$ is a cyclic group.

REMARK 4.21. *But the problem is: It is not easy to directly figure out or prove that there always exists in $K^\times$ an element of order $q-1$.*

To overcome this difficulty, we apply Proposition 4.18: Since $K^\times$ is a finite subgroup of $K^\times$ itself of order $q-1$, by Proposition 4.18 it is a cyclic group, consisting of all the $(q-1)$-th roots of unity in F, and this finishes the proof of part c.

Part d: Let K and K' be two fields of the same order q, and by part c, let α be a generator of the cyclic group $K^\times$. Then K is clearly generated as a field extension of $F = \mathbb{F}_p$ by this element α: $K = F(\alpha)$. Now let $f(x)$ be the irreducible polynomial for α over F. Then K is isomorphic to the field $F[x]/(f(x))$ by Proposition 4.6. On the other hand, since α is a root of two polynomials $f(x)$ and $x^q - x$, and since $f(x)$ is irreducible over F, $f(x)$ must divide $x^q - x$ in $F[x]$ by part e of Proposition 4.13. Thus since $x^q - x$ factors into linear factors in K' by part b, $f(x)$ has a root α' in K'. This means that we have

$$K \approx F[x]/(f(x)) \approx F(\alpha') \subseteq K'.$$

Finally, since the subfield $F(\alpha')$ and K' have the same order they must be the same; hence K and K' are isomorphic.

Part e: Let $f(x)$ be an irreducible polynomial of degree r in $F[x]$, where $F = \mathbb{F}_p$ as before. It has a root α in some field extension L of F, and the subfield $K := F(\alpha)$ of L has degree r over F by Proposition 4.10.

Thus K has order p^r, and by part b of the theorem, α is also a root of $x^q - x$, where $q = p^r$. Then since $f(x)$ is irreducible it must divides $x^q - x$ by part e of Proposition 4.13. Now, let $g(x)$ be an irreducible polynomial over F such that its degree k divides r. Then $r = ks$ for some s, and $g(x)$ divides $x^{q'} - x$ where $q' = p^k$, which in turn divides $x^q - x$ by Lemma 4.6. On the other hand, if $g(x)$ is irreducible and if its degree k does not divide r, then since $[K : F] = r$, $g(x)$ cannot have a root in K and hence $g(x)$ cannot divide $x^q - x$ by part b.

Part f: If k does not divide r, then $q = p^r$ is not a power of $q' = p^k$, and hence a field of order q cannot be an extension field of a field of order q'. On the other hand, if k does divide r then Lemma 4.6 and part a of the theorem show that the polynomial $x^{q'} - x$ has all roots in a field K of order q. Then part b of Proposition 4.17 says that K contains a subfield with q' elements.

This finishes the whole proof of Theorem 4.7. □

We note that part c of Theorem 4.7 says that we can list all non-zero elements of K as powers of a single suitably chosen one. But this is not obvious, even when we are given a prime field $\mathbb{F}_p$.

EXAMPLE 4.23. *Consider* $\mathbb{F}_7^\times$. *Since* $2^3 \equiv 1 (\bmod 7)$, 2 *is not its generator. But under* $\bmod 7$, *we have* $3^0 \equiv 1$, $3^1 \equiv 3$, $3^2 \equiv 2, 3^3 \equiv 6, 3^4 \equiv 4, 3^5 \equiv 5, 3^6 \equiv 1$. *Thus, the residue of* 3 *in* $\mathbb{F}_7^\times$ *is its generator:*

$$\mathbb{F}_7^\times = \{1, 3, 2, 6, 4, 5\}.$$

EXAMPLE 4.24. *Consider* $\mathbb{F}_{11}^\times$. *By the same method of trial and error used in the above example, we can check that the residue of* 2 *in* $\mathbb{F}_{11}^\times$ *is its generator:*

$$\mathbb{F}_{11}^\times = \{1, 2, 4, 8, 5, 10, 9, 7, 3, 6\}.$$

DEFINITION 4.11. *A generator for the cyclic group* $\mathbb{F}_p^\times$ *is called a* ***primitive element modulo*** p.

REMARK 4.22. *Again, we note that Theorem 4.7 does not tell us how to find a primitive element. It only says that there exists a primitive* *element. Which residues modulo p are primitive elements is not not well understood, but given a small prime p, we can easily find a primitive element by trial and error as we have seen in the above examples.*

Part e of Theorem 4.7 is a basis for many methods of factoring polynomials modulo p. We have a few examples when $q = 2^r$.

EXAMPLE 4.25. *Let $q = 2^2$. The field of order 4 is an extension field of degree 2 over $\mathbb{F}_2$, and hence its irreducible polynomial must have degrees either 1 or 2 over $\mathbb{F}_2$, because, by part e of the theorem, possible degrees are divisors $1, 2$ of the exponent 2 in 2^2. Thus in $\mathbb{F}_2[x]$, we must have an irreducible factorization of the form:*

$$x^4 - x = x(x-1)(x^2+?+1) \in \mathbb{F}_2[x],$$

where $x^2+?+1$ is not yet determined. On the other hand, we have

$$x^4 - x = x(x-1)(x^2+x+1) \in \mathbb{Z}[x],$$

where x^2+x+1 is irreducible over $\mathbb{Z}$ by Eisenstein. Thus,

$$x^4 - x \mod \mathbb{F}_2 = x^4 - x = x(x-1)(x^2+x+1) \in \mathbb{F}_2[x].$$

Thus, by equating this with the previous one,

$$x(x-1)(x^2+?+1) = x^4 - x = x(x-1)(x^2+x+1),$$

the irreducible factor $x^2+?+1$ of $x^4 - x$ over $\mathbb{F}_2$ we are looking for is in fact x^2+x+1, because $\mathbb{F}_2[x]$ is an integral domain.

EXAMPLE 4.26. *Let $q = 2^3$. The field of order 8 is an extension field of degree 3 over $\mathbb{F}_2$, and hence its irreducible polynomial over $\mathbb{F}_2$ must have degree 3. Moreover, degrees of irreducible factors of $x^8 - x$ are the divisors $1, 3$ of the exponent 3 in 2^3 by part e of the theorem; hence in $\mathbb{F}_2[x]$, $x^8 - x$ must have exactly two irreducible factors of degree 3 over $\mathbb{F}_2$. This is because: x, $x - 1$ are automatic irreducible factors of degree 1 of $x^8 - x$ over $\mathbb{F}_2$, and all irreducible factors of $x^8 - x$ must be distinct by part a of the theorem.*

On the other hand, we have

$$x^8 - x = x(x-1)(x^6 + x^5 + \cdots + x + 1) \in \mathbb{Z}[x],$$

where $x^6 + x^5 + \cdots + x + 1$ *is irreducible over* $\mathbb{Z}$ *by Eisenstein's Theorem (Theorem 2.11). But, its image under* mod 2 *must be:*

$$x^8 - x = x(x-1)(x^3 + ? + 1)(x^3 + ?? + 1) \in \mathbb{F}_2[x],$$

because the two irreducible polynomials of degree 3 *must divide* $x^8 - x$ *in* $\mathbb{F}_2[x]$*. This means that we must have an equality in* $\mathbb{F}_2[x]$*:*

$$x^6 + x^5 + \cdots + x + 1 = (x^3 + ? + 1)(x^3 + ?? + 1) \in \mathbb{F}_2[x].$$

Moreover, among polynomials of degree 3 *over* $\mathbb{F}_2$, x^3+x+1 *and* x^3+x^2+1 *are the only irreducible polynomials of degree* 3 *over* $\mathbb{F}_2$*. (See Example 2.16). Thus, we must have*

$$x^6 + x^5 + \cdots + x + 1 = (x^3 + x + 1)(x^3 + x^2 + 1) \in \mathbb{F}_2[x],$$

and hence

$$x^8 - x = x(x-1)(x^3 + x + 1)(x^3 + x^2 + 1) \in \mathbb{F}_2[x].$$

To compute in the field $\mathbb{F}_8$*, we choose a root* α *of one of these cubics, say of* $x^3 + x + 1$*. Then* $\{1, \alpha, \alpha^2\}$ *is an* $\mathbb{F}_2$*-basis of* $\mathbb{F}_8$*; hence*

$$\mathbb{F}_8 = \{0, 1, \alpha, 1+\alpha, \alpha^2, 1+\alpha^2, \alpha+\alpha^2, 1+\alpha+\alpha^2\}.$$

Computation in $\mathbb{F}_8$ *is done using the relation* $\alpha^3 + \alpha + 1 = 0$.

We note that the roots of $x^3 + x^2 + 1$ *must be among these* 8 *elements. We also note that* $\mathbb{F}_4$ *cannot be a subfield of* $\mathbb{F}_8$ *by part* f *of the theorem, because the exponent* 2 *in* 2^2 *does not divide the exponent* 3 *in* 2^3.

EXAMPLE 4.27. *Let* $q = 2^4$*. The field of order* 16 *is an extension field of degree* 4 *over* $\mathbb{F}_2$*, and hence its irreducible polynomial over* $\mathbb{F}_2$ *must have degree* 4*. Moreover, degrees of irreducible factors of* $x^{16} - x$ *are the divisors* $1, 2, 4$ *of the exponent* 4 *in* 2^4 *by part* e *of the theorem. Thus,* $x, x-1,$ *and* $x^2 + x + 1$ *are its factors, since the exponent* 2 *in* 2^2 *divides* *the exponent* 4 *in* 2^4 *so that, by part* f *of the theorem,* $\mathbb{F}_4$ *is a subfield of* $\mathbb{F}_{16}$.

On the other hand, we have an irreducible factorization of $x^{16} - x$ *in* $\mathbb{Z}[x]$*:*

$$\begin{aligned} x^{16} - x &= x(x-1)(x^2+x+1)(x^4+x^3+x^2+x+1) \\ &\quad \cdot(x^8 - x^7 + x^5 - x^4 + x^3 - x + 1). \end{aligned}$$

We obtain this as follows: Since $x^{16} - x$ *is divisible by* $x^3 - 1$ *and* $x^5 - 1$ *so* $x, x-1, x^2+x+1$ *and* $x^4+x^3+x^2+x+1$ *are its irreducible factors in* $\mathbb{Z}[x]$*, so by dividing it by these polynomials we obtain such a factorization. By trial and error one may check that the last polynomial of degree* 8 *is also irreducible in* $\mathbb{Z}[x]$*. In fact, by the following arguments, we only need to check whether it* $= g(x)h(x), \deg(g(x)) = \deg(h(x)) = 4$ *for some polynomials* $g(x)$ *and* $h(x)$*; hence the last polynomial is irreducible in* $\mathbb{Z}[x]$*:*

But the image of $x^{16} - x \pmod 2 = x^{16} - x$ *in* $\mathbb{F}_2[x]$ *can have only irreducible factors of degrees* $1, 2,$ *or* 4*, so that* $x^8-x^7+x^5-x^4+x^3-x+1$ *cannot be irreducible over* $\mathbb{F}_2$*. Then since all the irreducible factors of* $x^{16}-x$ *in* $\mathbb{F}_2[x]$ *must be distinct, we see that* $x^4+x^3+x^2+x+1$ *is irreducible over* $\mathbb{F}_2$*. This is because its possible irreducible factors of degree* $1, 2$ *are already shown up as factors of* $x^{16} - x$ *and, since* 3 *never divides the exponent* 4 *in* 2^4*, any irreducible polynomial of degree* 3 *in* $\mathbb{F}_2[x]$ *cannot be a factor of* $x^{16} - x$ *in* $\mathbb{F}_2[x]$*. A similar argument forces us to conclude that irreducible factors of* $x^8 - x^7 + x^5 - x^4 + x^3 - x + 1$ *in* $\mathbb{F}_2[x]$ *must be only those with degree* 4*:*

$$x^8 - x^7 + x^5 - x^4 + x^3 - x + 1 = (x^4 + x^3 + 1)(x^4 + x + 1).$$

For this factorization, we may also recall those lists in Example 2.16. Thus, we have in $\mathbb{F}_2[x]$*:*

$$x^{16} - x = x(x-1)(x^2+x+1)(x^4+x^3+x^2+x+1)(x^4+x^3+1)(x^4+x+1).$$

In this example we also note that $x^4 - x$ *is a factor of* $x^{16} - x$, *since* $\mathbb{F}_4$ *is a subfield of* $\mathbb{F}_{16}$.

Exercises for Section 4.6

EXERCISE 4.14. *Explain why* $\mathbb{F}_4$ *is not isomorphic to* $\mathbb{Z}/4\mathbb{Z}$.

EXERCISE 4.15. *Determine all the irreducible polynomials of degree* 3 *over* $\mathbb{F}_3$.

EXERCISE 4.16. *Let* K *be a finite field. Prove that the product of the non-zero elements of* K *is equal to* -1.

EXERCISE 4.17. *Prove that every element of* $\mathbb{F}_p$ *has exactly one pth root.*

CHAPTER 5

Galois Theory

We first briefly recall the history of the solution of equations by radicals. The solution of the quadratic was known to the Hindus and in its geometric form to the Greeks. The cubic and quartic were solved by the Renaissance Italian mathematicians Scipio del Ferro (1515) and Ferrari (1545). However, not until the19th century did Abel (1802 - 1829) and Galois (1811 - 1832) prove the impossibility of solving all polynomial equations of degree $n \geq 5$.

History also tells us that Johann Carl Friedrich Gauss's $(1777-1855)$ doctoral dissertation was a discussion of the fundamental theorem of algebra in 1799 and Evariste Galois lived during $1811-1832$, so we now easily assume that he knew about the mathematical usefulness of complex numbers. Thus, all throughout this Chapter, we always assume that the characteristic of the bottom field in a field extension is 0 (i.e., ∞).

On the other hand, the origins and history of finite fields can be traced back to the 17th and 18th centuries, but there, it is known that these fields played only a minor role in the mathematics of the day. In fact, it was Galois who proved important results in the theory of finite fields or a Galois fields (that is, Theorem 4.7), which requires the theory of Groups, a new theory not yet well developed until that time.

This should be the case, because history also tells us that Galois was the first mathematician who recognized the importance of symmetry on the set of mathematical objects, and it was he who provided concrete examples and consequently this urged others to the study on the group theory. It is known that Dedekind $(1831-1916)$, the last Ph.D. student of Gauss, coined the official definition of a group.

1. Primitive Elements in Characteristic 0 Case

Assuming that every field under consideration in this chapter is of characteristic zero will allow us to understand the basics of the Galois theory quite easily.

In the last chapter, following Kronecker's construction of a new field, the so-called symbolic adjunction of roots, we studied algebraic field extensions, using extensions generated by a single element as the basic tool. This amounts to studying the properties of a single root of an irreducible polynomial

$$f(x) = x^n + a_{n-1}x^{n-1} + \cdots + a_0.$$

Galois theory, the topic of this chapter, is about the theory of all the roots of such a polynomial and of the symmetries among them.

NOTATION 5.1. *The notation*

$$K/F$$

will always mean that K is an extension field of F.

As we have seen in the last Chapter, computation in a field $F(\alpha)$ generated by a single root α of $f(x) = 0$ can easily be made by identifying it with the formally constructed field $F[x]/(f(x))$.

We now suppose that $f(x) = 0$ has n (distinct) roots in an extension field K of F:

$$\alpha_1, \alpha_2, ..., \alpha_n.$$

Then we have

$$f(x) = (x - \alpha_1) \cdots (x - \alpha_n),$$

and by comparing the expansion of this expression with the original expression for $f(x)$, we can easily see each a_j is a polynomial expression of the α_i's. But it is not easy to express each α_j as a polynomial of the a_i's.

The fundamental discovery which arose through the work of several mathematicians, especially of Lagrange and Galois, is that the relationship between the roots can be understood in terms of symmetry. The original model for this symmetry is complex conjugation, which permutes the two roots $\pm i$ of the irreducible real polynomial $x^2 + 1$, while leaving the real numbers fixed. As we will see the usefulness of complex conjugation in Section 6 of this Chapter, the significance of complex conjugation is remarkable.

In this section, we are first going to introduce several definitions for Galois Theory.

DEFINITION 5.1. *An F-**automorphism** of an extension field K is an automorphism, i.e., an isomorphism: $K \longrightarrow K$, which is the identity map on the subfield F.*

Thus, an automorphism σ of K is an F-automorphism if $\sigma(c) = c$ for each $c \in F$, the bottom field.

EXAMPLE 5.1. *Complex conjugation is an $\mathbb{R}$-automorphism of $\mathbb{C}$.*

As we shall see in later sections on Equations, symmetries are all about the behavior of F-automorphisms, and it needs more terminologies for a precise description of the theory.

DEFINITION 5.2. *The group of all F-automorphisms of K is called the **Galois group** of the field extension.*

NOTATION 5.2. *We usually denote the Galois group by*

$$Gal(K/F).$$

One easily verifies that

COROLLARY 5.1. *$Gal(K/F)$ is a group.* $\square$

DEFINITION 5.3. *Let $f(x) \in F[x]$ be a non-constant polynomial. A **splitting field** for $f(x)$ over F is an extension field K of F such that*

(1) *$f(x) = (x - \alpha_1) \cdots (x - \alpha_n)$ in $K[x]$ with each $\alpha_i \in K$.*
(2) *$K = F(\alpha_1, ..., \alpha_n)$.*

That is, K is a splitting field for $f(x)$ if $f(x)$ factors into linear polynomials over K and $K[x]$ is the smallest field extension of F that contains all the roots of $f(x) = 0$. We note that

PROPOSITION 5.1. *Every polynomial $f(x) \in F[x]$ has a splitting field.*

PROOF. By Proposition 4.12, we have a field extension L in which $f(x)$ splits into linear factors. We then take for K the subfield $F(\alpha_1, ..., \alpha_n)$ of L generated by the roots, $\alpha_1, ..., \alpha_n$ of $f(x) = 0$. □

PROPOSITION 5.2.

- **a:** Let K be an extension field of a field F, $f(x)$ a polynomial in $F[x]$, and let σ be an F-automorphism of K. If α is a root of $f(x) = 0$ in K, then so is $\sigma(\alpha)$.
- **b:** Let $K = F(\alpha_1, ..., \alpha_n)$ be a field extension generated over F by $\alpha_1, ..., \alpha_n$, and let σ be an F-automorphism of K. If σ fixes each of the generators $\alpha_1, ..., \alpha_n$, then σ is the identity map: $K \longrightarrow K$.
- **c:** Let K be a splitting field of a polynomial $f(x)$ over F. The Galois group $G(K/F)$ operates *faithfully* on the set

$$S = \{\alpha_1, ..., \alpha_n\}$$

of roots in K of $f(x) = 0$, that is, the homomorphism of groups: $G(K/F) \longrightarrow Perm(S)$ is *injective.*

REMARK 5.1. *One recalls that if a group G acts on a set S, then there eixsts an associated group homomorphism: $G \longrightarrow Perm(S)$, and conversely. In Part c of Proposition 5.2, it is the situation that the group $G(K/F)$ acting on K also acts automatically on the subset $S = \{\alpha_1, ..., \alpha_n\}$ of K.*

PROOF. We have already proved part a in Proposition 4.8.

To prove part b, we assume that $K = F(\alpha_1, ..., \alpha_n)$. Thus, every element of K can be expressed as a polynomial in $\alpha_1, ..., \alpha_n$ with coefficients in F. Thus, if σ fixes each element α_i and if it is the identity map on F, then it fixes every polynomial in $\{\alpha_1, ..., \alpha_n\}$ over F; hence it must be the identity map: $K \longrightarrow K$.

Part c follows from the first two, and it goes as follows: Part a tells us that every $\sigma \in G(K/F)$ permutes the elements of the set $\{\alpha_1, ..., \alpha_n\}$, and then part b tells us that the operation on this set is faithful. □

REMARK 5.2. *We note that Proposition 5.2 does not describe which permutations of the roots of a polynomial equation extend to automorphisms of the splitting field.*

The following example is the most simplest one which connect the theory of finite groups with the theory of finite field extension.

EXAMPLE 5.2. *Complex conjugation σ sending $a + bi \mapsto a - bi$ where $i = \sqrt{-1}$ is an $\mathbb{R}$-automorphism of $\mathbb{C}$. Since $\sigma^2 =$ the identity map, and hence by Proposition 5.2, $Gal(\mathbb{C}/\mathbb{R})$ is a cyclic group consisting of two $\mathbb{R}$-automorphisms, the identity map and complex conjugation. Moreover, we have in this case*

$$|Gal(\mathbb{C}/\mathbb{R})| = 2 = [\mathbb{C} : \mathbb{R}] .$$

REMARK 5.3. *Very soon in this section, we are going to prove that if K/F is a finite field extension, then the order, i.e., the cardinality*

$$|Gal(K/F)|$$

of $Gal(K/F)$ divides the degree $[K : F]$ of the extension. (Theorem 5.3).

DEFINITION 5.4. *If G is a group of automorphisms of a field K, the set of elements of K which are fixed by all the automorphisms in G is denoted by*

$$K^G .$$

That is, $K^G = \{\alpha \in K : \varphi(\alpha) = \alpha \text{ for all } \varphi \in G\}$.

REMARK 5.4. *In other words we have the following observation: Let G be the set of all automorphisms of a field K. Then G is clearly a group acting on K as is explained now: For any $g \in G$ and any $a \in K$, we define*

$$g.a := g(a).$$

Then one can easily verify that this gives a group action of G on K. Under this action, the field K is partitioned into a union of disjoint orbits $\mathcal{O}_a$ of various elements a of K. Among these orbits, 1 in K is always fixed under the group action; hence its orbit consists of itself alone. Thus, *there are generally more than one* "singleton" *orbit and consequently the subset* K^G *of* K *is nothing but the union of these djsjoint singleton orbits.*

On the other hand, $Gal(K/F)$ *is clearly a subgroup of* G*.*

One then easily verifies that

COROLLARY 5.2. K^G *is a subfield of* K*, called the* ***fixed field*** *of* G*.*

□

DEFINITION 5.5. *An element* γ *which generates a field extension* K/F *is called a* **primitive element** *for* K *over* F*.*

THEOREM 5.1. *(Existence of a primitive element) Let* K *be a finite extension field of a field* F *of characteristic* 0*. Then there exists an element* $\gamma \in K$ *such that* $K = F(\gamma)$*.*

REMARK 5.5. *This theorem is not true for fields of characteristic* p*, where* p *is a prime number* > 1*.*

PROOF. By hypothesis K is finite over F, so $K = F(\alpha_1, ..., \alpha_n)$. We use induction on the number n of generators α_i's of K over F. If $n = 1$, there is nothing to prove. For $n > 1$, by induction hypothesis we assume that the theorem is true for the intermediate field $K_1 = F(\alpha_1, ..., \alpha_{n-1})$. Thus, $K_1 = F(\beta)$ for some $\beta \in K_1$; hence $K = K_1(\alpha_n) = F(\beta, \alpha_n)$. Thus, we are reduced to the case when $n = 2$, so that K is generated over F by two elements α, β.

Thus, we need to prove that this $K = F(\alpha, \beta)$ is generated by a single element over F:

First of all, since any finite extension is an algebraic extension by Corollary 4.3, we know that α and β are algebraic over F. Thus we let $f(x), g(x)$ be the irreducible polynomials for α, β over F, respectively. Then by Proposition 4.12 we have an extension field K' of $K = F(\alpha, \beta)$ where the product $f(x) \cdot g(x)$ splits completely; hence in which $f(x)$ and $g(x)$ split completely. Let $\alpha = \alpha_1, \alpha_2, ..., \alpha_n$ be roots in K' of $f(x) = 0$ and $\beta = \beta_1, \beta_2, ..., \beta_m$ be roots in K' of $g(x) = 0$. Then by Proposition

4.15 these roots are all distinct because the characteristic of F is 0 by hypothesis.

We are now going to show that for an undetermined coefficient $c \in F$ (which is going to be determined later), the linear combination

$$\gamma = \beta + c\alpha$$

generates $K = F(\alpha, \beta)$:

Obviously we have $F(\gamma) \subseteq K = F(\alpha, \beta)$, and hence it is sufficient to show that $\alpha \in F(\gamma)$, because if so, then $\beta = \gamma - c\alpha \in F(\gamma)$.

In order to show that $\alpha \in F(\gamma)$, we are going to determine its irreducible polynomial over $F(\gamma)$ and it goes as follows. We know that α is a root of $f(x) = 0$. Consider

$$h(x) := g(\gamma - cx).$$

It is clearly a polynomial over $F(\gamma)$, and $h(\alpha) = 0$. Thus, if we can show that the greatest common divisor of $f(x)$ and $h(x)$ in $F(\gamma)[x]$ is $x - \alpha$, then it will follow that $-\alpha$, the constant term of $x - \alpha$, is an element of $F(\gamma)$:

For this, we recall that the monic greatest common divisor of $f(x)$ and $h(x)$ is the same, whether it is computed in $F(\gamma)[x]$ or $K'[x]$ by Proposition 4.13. Thus, we may make our computation in $K'[x]$. In that polynomial ring, $f(x)$ is linearly factored:

$$f(x) = (x - \alpha)(x - \alpha_2) \cdots (x - \alpha_n) \in K'[x],$$

and hence it is sufficient to show that none of $x - \alpha_i$, $i = 2, ..., n$, except for $\alpha = \alpha_1$, divides $h(x)$ in $K'[x]$. For this, we need to figure out all the roots in K' of $h(x) = 0$ and it goes as follows: Since the β_j's are roots in K' of $g(x) = 0$, the roots of $h(x) = g(\gamma - cx) = 0$ are determined by solving the equations for x:

$$\gamma - cx = \beta_j.$$

Then since $\gamma = \beta + c\alpha$ holds, the roots are

$$\frac{\gamma - \beta_j}{c} = \frac{\beta - \beta_j}{c} + \alpha.$$

Since we are looking for a reasonable condition on c under which

$$\gcd(f(x), h(x)) = x - \alpha$$

holds, we want these roots to be different from each α_i, for $i = 2, ..., n$, and this will be the case *provided that* c is not one of the finitely many values,

$$-\frac{\beta_j - \beta}{\alpha_i - \alpha}, \tag{5.1}$$

with $i = 2, ..., n$ and $j = 2, ..., m$. □

REMARK 5.6. *Since finitely many inadmissible values for $c \in F$ are explicitly described, this proof is very constructive in nature, and consequently, it also shows us that there should be many primitive elements γ in general.*

EXAMPLE 5.3. *Let K be the field $\mathbb{Q}[i, \sqrt[3]{2}]$. This field has degree 6 over $\mathbb{Q}$, because no smaller numbers than 6 are divisible by both by 2 and 3. For a similar argument, see Example 4.9. Adapting the same notation used in the previous proof, $\beta = \beta_1 = i, \beta_2 = -i$, and $\alpha = \alpha_1 = \sqrt[3]{2}, \alpha_2 = \omega\sqrt[3]{2}$, and $\alpha_3 = \omega^2\sqrt[3]{2}$, where $\omega = e^{2\pi i/3}$. Then condition (5.1) becomes*

$$\sqrt[3]{2}c \neq -\frac{\pm i - 1}{\omega^\nu - 1}, \qquad \nu = 1, 2.$$

We note that this condition holds for all $c \in \mathbb{Q}$ except for $c = 0$. Therefore, $\gamma = i + c\sqrt[3]{2}$ generates K over $\mathbb{Q}$ for all rational numbers $c \neq 0$.

EXAMPLE 5.4. *(This is the continuation of Example 5.3.) Of course, many other combinations of the two elements α, β will generate $K = F[\alpha, \beta]$. In the case of this example, the product $i\sqrt[3]{2}$ also generates K, because its irreducible polynomial over $\mathbb{Q}$ is $x^6 + 4$. This is very easy since $(i\sqrt[3]{2})^2 = 2^{2/3}, (i\sqrt[3]{2})^3 = -2i, (i\sqrt[3]{2})^6 = -4$, and hence,*

$$x^6 + 4$$

is obviously the minimal polynomial over $\mathbb{Q}$ *of the product* $i\sqrt[3]{2}$, *since no other polynomials in* $\mathbb{Q}[x]$ *of less degree than* 6 *can have* $i\sqrt[3]{2}$ *as its root.*

REMARK 5.7. *Therefore, Theorem 5.1 is very important for at least two reasons.*

First, explicit computation in an extension of the form $F(\gamma)$ *is very easy if the irreducible polynomial for* γ *over* F *is known to us or computable.*

Second, since finite extensions have the form $F(\gamma)$, *we can derive their properties from facts about algebraic elements. The second aspect of Theorem 5.1 is most important for us.*

REMARK 5.8. *But, as for the first reason, the next two examples suggest that there is a lot more general and powerful method based on the theory of Groebner Basis.*

EXAMPLE 5.5. *(This is the continuation of Example 5.3 and Example 5.4.) We now know that any* $\gamma = i + c\sqrt[3]{2}$ *with a non-zero* $c \in \mathbb{Q}$ *is a primitive element for* $K = \mathbb{Q}(i, \sqrt[3]{2})$. *Then, for instance, what is the minimal polynomial over* $\mathbb{Q}$ *of the primitive element* $i + \sqrt[3]{2}$*?*

The minimal polynomial of $\sqrt[3]{2}$ *over* $\mathbb{Q}$ *is clearly* $x^3 - 2$, *and since* $\mathbb{Q}(\sqrt[3]{2})$ *is a subfield of* $\mathbb{R}$, *the minimal polynomial* $y^2 + 1$ *of* $i = \sqrt{-1}$ *over* $\mathbb{R}$ *is again the minimal polynomial of* i *over* $\mathbb{Q}(\sqrt[3]{2})$.

Then the following codes for a CAS(Computer Algebra System), REDUCE, will provide the right answer very quickly:

```
load_package groebner$
off nat$
q1:=x^3-2$
q2:=y^2+1$
q3:=z-(x+y)$
result1:=groebner({q1,q2,q3},{x,y,z});
```

*result1:={22*x - 12*z**5 + 9*z**4 - 40*z**3 + 78*z**2 - 100*z - 91, 22*y + 12*z**5 - 9*z**4 + 40*z**3 - 78*z**2 + 78*z + 91,* ***z**6 + 3*z**4 - 4*z**3 + 3*z**2 + 12*z + 5****}$*

———————

Thus, the last equation in result1 with only variable z, changing z to x for our convenience,

$$x^6 + 3x^4 - 4x^3 + 3x^2 + 12x + 5,$$

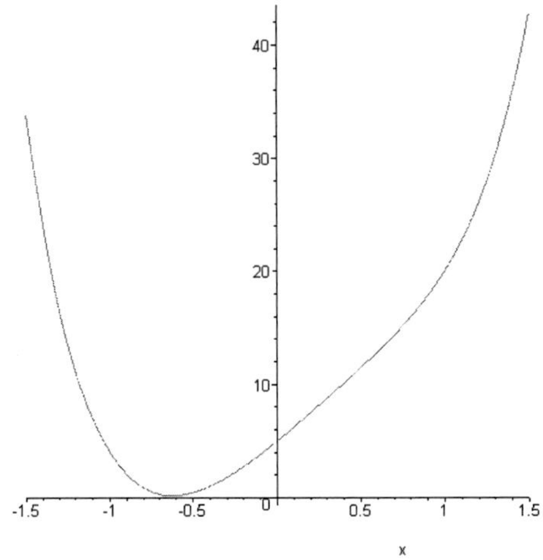

is the minimal polynomial over $\mathbb{Q}$ *of* $i+\sqrt[3]{2}$*, i.e., the irreducible polynomial over* $\mathbb{Q}$ *of least degree of* $i+\sqrt[3]{2}$.

EXAMPLE 5.6. *If we examine the computed reduced Groebner basis of Example 5.5 above, we immediately see that* $i+\sqrt[3]{2}$ *is a primitive element for* $\mathbb{Q}(i, \sqrt[3]{2})$ *over* $\mathbb{Q}$*, and*

$$\begin{aligned} \sqrt[3]{2} &= \frac{-1}{22}\left(\begin{array}{c} -12\left(i+\sqrt[3]{2}\right)^5 + 9\left(i+\sqrt[3]{2}\right)^4 - 40\left(i+\sqrt[3]{2}\right)^3 \\ +78\left(i+\sqrt[3]{2}\right)^2 - 100(i+\sqrt[3]{2}) - 91 \end{array}\right), \\ i &= \frac{-1}{22}\left(\begin{array}{c} 12\left(i+\sqrt[3]{2}\right)^5 - 9\left(i+\sqrt[3]{2}\right)^4 + 40\left(i+\sqrt[3]{2}\right)^3 \\ -78\left(i+\sqrt[3]{2}\right)^2 + 78\left(i+\sqrt[3]{2}\right) + 91 \end{array}\right). \end{aligned}$$

Here are the actual codes for such computation for the REDUCE.

———————

```
off echo$
off nat$
```

```
f := ((-1)/22)*( - 12*z**5 + 9*z**4 - 40*z**3 + 78*z**2 - 100*z -
91);
g:=((-1)/22)*(12*z**5 - 9*z**4 + 40*z**3 - 78*z**2 + 78*z + 91);
result1:= sub({z=i+2^(1/3)},f);
2**(1/3)$
result2:= sub({z=i+2^(1/3)},g);
i$
```

——————————

This means that there is an important theory behind this computation, the theory of "Groebner basis".

EXAMPLE 5.7. *This is the continuation of Example 5.5. More generally,*

——————————

```
off echo$
off nat$
load_package groebner$
q1:=x^3-2$
q2:=y^2+1$
q3:=z-(c*x+y)$
result1:=groebner({q1,q2,q3},{x,y,z});
result1 := {(54*c**7 - 32*c)*x - 12*z**5 + (9*c**3)*z**4 - 40*z**3
+ (78*c**3)*z**2 - (72*c**6 + 28)*z - 91*c**3, (54*c**6 - 32)*y +
12*z**5 - (9*c**3)*z**4 + 40*z**3 - (78*c**3)*z**2 + (18*c**6+ 60)*z
+ 91*c**3, z**6 + 3*z**4 - (4*c**3)*z**3 + 3*z**2 + (12*c**3)*z +
(4*c**6 + 1)}$
```

——————————

Thus, the last equation in result1 with only variable z,

$$x^6 + 3x^4 - 4c^3x^3 + 3x^2 + 12c^3x + 4c^6 + 1$$

is the minimal polynomial over $\mathbb{Q}$ of $i + c\sqrt[3]{2}$, i.e., the irreducible polynomial over $\mathbb{Q}$ of least degree of $i + c\sqrt[3]{2}$, for a non-zero $c \in \mathbb{Q}$.

We also note that when $c = 1$*, it agrees well with the result in Example 5.5, which should be the case.*

REMARK 5.9. *The above examples certainly require a good deal of the theory of Groebner Basis, which is beyond the scope of this book. In our other book, which is a continuation of this book, we will study the theory of Groebner Basis and its applications in detail.*

If you understand the meaning of the assignment for q3 in the above program, then you can use this program for other situations as well. That is, if $K = F(\alpha_1, ..., \alpha_n)$ *and if*

$$\gamma = \frac{f(\alpha_1, ..., \alpha_n)}{g(\alpha_1, ...\alpha_n)} \in K,$$

with $f(x_1, ..., x_n), g(x_1, ..., x_n) \in F[x_1, ..., x_n]$*, then*

$$q3 := g(x_1, ..., x_n)x_{n+1} - f(x_1, ..., x_n)$$

is the needed expression.

In the case of the above example, $\alpha_1 = i$ *and* $\alpha_2 = \sqrt[3]{2}$*, and* $i + c\sqrt[3]{2}$ *is a polynomial in* i *and* $\sqrt[3]{2}$ *with coefficients* 1 *and* $c \in \mathbb{Q}$*. Thus,* $g(x, y) \equiv 1$ *and* $f(x, y) = cx + y$*.*

REMARK 5.10. *As the following example shows, we can still compute the minimal polynomial for*

$$\beta = \frac{p(\alpha_1, ..., \alpha_n)}{q(\alpha_1, ..., \alpha_n)}$$

even if we are only given those n *polynomials* $h_i(x)$ *whose roots are* α_i *for each* $i = 1, ..., n$*.*

EXAMPLE 5.8. *This is a continuation of Example 5.5, and here in this example we replace the previous q1:=x^3-2 and q2:=y^2+1 with*

$$q1 \ : \ = (x\hat{\ }2 + x + 3) * (x\hat{\ }3 - 2),$$
$$q2 \ : \ = (y\hat{\ }2 - 5) * (y\hat{\ }2 + 1),$$

respectively, and try to compute the minimal polynomial of $i + \sqrt[3]{2}$ *over* $\mathbb{Q}$ *by examining factors of the single computed Groebner basis element for the ideal generated by the minimal polynomial of* $i + \sqrt[3]{2}$ *over* $\mathbb{Q}$. *For this purpose, we use the Computer Algebra System, REDUCE:*

```
————————
off echo$
off nat$
load_package groebner$
q1:=(x^2+x+3)*(x^3-2)$
q2:=(y^2-5)*(y^2+1)$
q3:=z-(x+y)$
gb:=groebner({q1,q2,q3},{x,y,z});
gb := {504894319888364027596874269962900*x
+ 30952587292283330128066172724*z**19
+ 1019822120177303427429391676 83*z**18
- 133111019967260750728011892994*z**17
- 114345202558468165178634177964 2*z**16
- 149282676819814715677999102801 2*z**15
- 119494338933330440970814716071 9*z**14
+ 78761209500434649111274534940 02*z**13
- 224290363016785337201671709501 86*z**12
- 184573475373052854675305769748372*z**11
- 223666069980306259653843349498034*z**10
+ 323328951421891594176856616528896*z**9
- 389175831927428247706848234280618*z**8
+ 482489260304885965628816085644336*z**7
- 868554897692487985359724441516118 3*z**6
- 243360605274204112731238939699616 6*z**5
+ 848937212901837561748557390916077 2*z**4
- 449470499840564390094265017004600 4*z**3
- 230315305528869990738090133545718 83*z**2
- 247802112430780999950932450691126 50*z
- 836489028980527271707532097372515 0,
504894319888364027596874269962900*y
```

```
- 30952587292283330128066172724*z**19
- 101982212017730342742939167683*z**18
+ 133111019967260750728011892994*z**17
+ 1143452025584681651786341779642*z**16
+ 1492826768198147156779991028012*z**15
+ 1194943389333304409708147160719*z**14
- 7876120950043464911127453494002*z**13
+ 22429036301678533720167170950186*z**12
+ 184573475373052854675305769748372*z**11
+ 223666069980306259653843349498034*z**10
- 323328951421891594176856616528896*z**9
+ 389175831927428247706848234280618*z**8
- 482489260304885965628816085644336*z**7
+ 8685548976924879853597244415161183*z**6
+ 2433606052742041127312389396996166*z**5
- 8489372129018375617485573909160772*z**4
+ 4494704998405643900942650170046004*z**3
+ 23031530552886999073809013354571883*z**2
+ 24275316923189735967496370799149750*z
+ 8364890289805272717075320973725150,
z**20 + 4*z**19 - 2*z**18 - 40*z**17 - 74*z**16 - 72*z**15
+ 226*z**14 - 548*z**13
- 6481*z**12 - 11392*z**11 + 5482*z**10 - 5324*z**9 +
6065*z**8 - 268504*z**7
- 275738*z**6 + 228660*z**5 + 39940*z**4 - 859844*z**3
- 1293584*z**2
- 790300*z - 178475}$
result:=factorize(part(gb,length(gb)));
result := {{z**6 + 3*z**4 - 4*z**3 + 3*z**2 + 12*z + 5,1},
{z**6 - 15*z**4 - 4*z**3 + 75*z**2 - 60*z - 121,1},
{z**4 + 2*z**3 + 9*z**2 + 8*z + 5,1},
{z**4 + 2*z**3 - 3*z**2 - 4*z + 59,1}}$
```

```
on rounded$
j:=1$
while j leq length(result) do
begin
write "result",j," :=",part(sub({z=i+2^(1/3)},part(result,j)),1);
j:=j+1$
end$
```

```
result1 :=0$
result2 :=56.1429467773*i - 66.3392840344$
result3 :=41.1633010103*i + 10.8018873555$
result4 := - 1.07480418714*i + 42.6340221331$
```

————————

Thus, we conclude from the above computation that

$$z^6 + 3z^4 - 4z^3 + 3z^2 + 12z + 5$$

is indeed the minimal polynomial of $i + \sqrt[3]{2}$ over $\mathbb{Q}$.

REMARK 5.11. *As in the above example, even if we are going to use the algorithm for factorization of polynomials, it would be a lot better if we start our computation with all irreducible polynomial inputs. That is, if we start our computation with those irreducible polynomials h_i over the bottom field k each of which has the given α_i as a root, then it will certainly reduce computation time a lot.*

Even in this case, it could happen that some of the h_i's are not irreducible over the intermediate fields properly containing k, so that we will still need the codes in Example 5.8.

For example, if we take

$$q1 := (x - i) * (x\hat{\ }3 - 2)$$

instead of the q1:=(x^2+x+3)(x^3-2) in the above example, then this new q1 is not irreducible over the intermediate field $\mathbb{Q}(i)$ and the codes in Example 5.8 certainly compute the minimal polynomial of $i + \sqrt[3]{2}$ over $\mathbb{Q}$.*

The strength of Theorem 5.1 is measured by applying it to the study of automorphisms of fields. Thus, in what follows, after having a few results, we are going to prove Theorem 5.3.

PROPOSITION 5.3. *Let G be a* finite *group of order n of automorphisms of a field K. Let $\{\beta = \beta_1, \beta_2, ..., \beta_d\}$ be the orbit of an element $\beta = \beta_1 \in K$ under the action of the group G. Then β is algebraic over its fixed field K^G, its degree over K^G is d, and its irreducible polynomial over K^G is*

$$g(x) = (x - \beta)(x - \beta_2) \cdots (x - \beta_d).$$

Moreover, the degree of β divides the order of the group G.

REMARK 5.12. *To have a better understanding of Proposition 5.3, one may refer to Remark 5.4. Namely, since K^G is the collection of all disjoint singleton orbits under the action of the given finite group G, the order d of any non-singleton orbit $\mathcal{O}_\beta$ for any $\beta \notin K^G$ is finite > 1, and it divides the order n of G.*

PROOF. Write $L = K^G$. Then any element of G is automatically an L-automorphism.

Now, we consider the polynomial:

$$g(x) = (x - \beta)(x - \beta_2) \cdots (x - \beta_d).$$

Since $g(x)$ is fixed under all permutations of $\{\beta = \beta_1, \beta_2, ..., \beta_d\}$, and hence by the operation of G, its coefficients are all in $L = K^G$. This means that $g(x) \in L[x]$. Now, let $f(x)$ be the irreducible polynomial for β over $L = K^G$. By part a of Proposition 5.2, each β_i is a root of $f(x) = 0$; thus $g(x)$ divides $f(x)$ in the splitting field of $f(x)$ over L. Thus, $g(x)$ divides $f(x)$ in $L[x]$ by part b of Proposition 4.13. Then since $f(x)$ is irreducible over L, we must have $f(x) = g(x)$ in $L[x]$. Finally, the degree of β divides the order of G because it is the order of an orbit under the action of G: $|G| = |G_\beta|\,|\mathcal{O}_\beta|$. □

REMARK 5.13. *Under such a hypothesis of the above proposition, this proposition provides a method for determining the irreducible polynomial for an element β of a Galois extension K over F (Definition 5.6).*

EXAMPLE 5.9. *We consider an easy example: $\beta = i + \sqrt{2} \in K = \mathbb{Q}(i, \sqrt{2})$. The Galois group of $K/\mathbb{Q}$ consists of 4 elements,*

$$\begin{aligned}
&\sigma\colon a + bi \mapsto a - bi,\\
&\tau\colon a + b\sqrt{2} \mapsto a - b\sqrt{2},\\
&\sigma\tau = \tau\sigma,\\
&\text{the identity map } 1.
\end{aligned}$$

[We will see the details of this in a later section, Section 6 of this Chapter. Biquadratic Equations]. Then since $\sigma^2 = 1$, $\tau^2 = 1$, and $(\sigma\tau)^2 = 1$, it is the Klein 4 group, and the orbit of β consists of the four elements $\pm i \pm \sqrt{2}$. Thus by Proposition 5.1, the irreducible polynomial for β over $\mathbb{Q}$ is

$$(x - (i + \sqrt{2}))(x - (i - \sqrt{2}))(x - (-i + \sqrt{2}))(x - (-i - \sqrt{2})).$$

If we expand it, it becomes $(x^2 - 2ix - 3)(x^2 + 2ix - 3) = x^4 - 2x^2 + 9$.

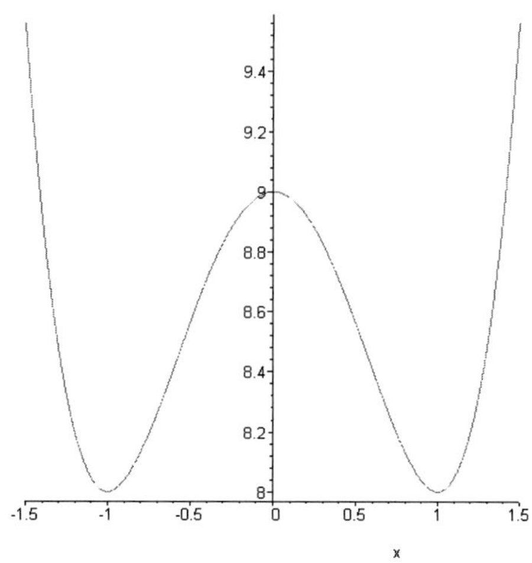

REMARK 5.14. *We put three remarks here.*

(1) *We can also determine this polynomial by computing powers of β and finding the linear relation of smallest degree between them,*

as we have seen in Example 4.11. However, this method of undetermined coefficients does not always produce the irreducible polynomial for β over the bottom field F.

(2) *The method described in Proposition 5.3 seems better than the method of undetermined coefficients, but under the condition that we already know what constitutes all the F-automorphisms of K/F.*

(3) *On the other hand, we have seen in Example 5.5 a general method of determining the irreducible polynomial for any γ in $F(\alpha_1, ..., \alpha_n)$ especially when $n = 2$. It is an application of the theory of Groebner Basis.*

Exercises for Section 5.1

EXERCISE 5.1. *Let G be a group of automorphisms of a field K. Prove that the fixed elements K^G is a subfield of K.*

EXERCISE 5.2. *Let $\alpha = \sqrt[3]{2}$, $\omega = \frac{1}{2}(-1+\sqrt{-3})$, and $\beta = \alpha\omega$.*

(1) *Prove that for all $c \in \mathbb{Q}, \gamma = \alpha + c\beta$ is the root of a 6th-degree polynomial of the form*

$$x^6 + ax^3 + b.$$

(2) *Prove that the irreducible polynomial for $\alpha + \beta$ is cubic.*

(3) *Prove that $\alpha - \beta$ has degree 6 over $\mathbb{Q}$.*

EXERCISE 5.3. *Using Computer Algebra System, for instance REDUCE, verify the results of the previous Exercise.*

2. Automorphisms of a Field

THEOREM 5.2. *Let G be a* finite *group of order n of automorphisms of a field K. Then $\left[K : K^G\right] = n$.*

PROOF. By Proposition 5.3, every element β of K is algebraic over its fixed field $L := K^G$, and moreover its degree divides $n = |G|$, and hence it is upper bounded by n. This is because the degree of the irreducible polynomial for β is the order of the orbit containing β under the group action of G on K; hence it divides $|G|$.

CLAIM 5.1. $[K : L] \leq n$.

PROOF OF CLAIM. Let α be an element of K such that the degree $[L(\alpha) : L]$ is maximal, say $m \leq n$. If $L(\alpha) \neq K$, then there exists an element $\beta \in K - L(\alpha)$ such that $L1 := L(\alpha)$ is *properly* contained in $L_2 := L_1(\beta) = L(\alpha, \beta)$. Then since both α and β are algebraic over L, $[L_2 : L] = [L_2 : L(\alpha)] \cdot [L(\alpha) : L]$ is finite, thus by Theorem 5.1, there exists $\gamma \in K$ such that $L_2 = L(\gamma)$. Thus again by Proposition 5.3 γ is algebraic over L and the degree of γ over $L := K^G$ is bounded by n. But, the chain

$$L \subset L(\alpha) \subsetneqq L(\alpha, \beta)$$

clearly means that $[L_2 : L] > m$, and hence γ has degree $> m$, which contradicts to our choice of α. Therefore, we must have $L(\alpha) = K$ and hence $[K : L] \leq n$, and this finishes the proof of Claim. □

Thus again by Theorem 5.1, we see that K has a primitive element γ over L: $K = L(\gamma)$. Then since G is a group of L-automorphisms of K, any element of G which fixes γ is in fact the identity map on $K = L(\gamma)$ by part b of Proposition 5.2. This means that the stabilizer of γ, which is a subgroup of G, consisting only of the identity map, is of order 1, and hence the orbit containing γ under the group action of G on K must have order $n = |G|$, because $n = 1 \cdot n$. Then by Proposition 5.3, we know that γ has degree n over $L = K^G$, and hence $\left[K : K^G\right] = n$, since $K = L(\gamma)$. □

REMARK 5.15. *We note that the arguments in the above proof are based on the counting formula in Group Theory, which says that "(the size of the orbit containing γ) times (the size of stabilizer of γ) is equal to the order of G."*

THEOREM 5.3. *For any* finite *field extension K/F, the order $|G(K/F)|$ of the Galois group $G(K/F)$ divides the degree $[K : F]$ of the extension.*

PROOF. Let $G = G(K/F)$. Then G operates on K, and hence by Proposition 5.2 and Theorem 5.2, we have $|G| = \left[K : K^G\right]$. Then since we have

$$F \subset K^G \subset K,$$

$\left[K : K^G\right]$ divides $[K : F]$ by Theorem 4.3. □

DEFINITION 5.6. *A finite field extension K/F is called a* ***Galois extension*** *if the order of the Galois group is equal to the degree:*

$$|Gal(K/F)| = [K : F] .$$

COROLLARY 5.3. *Let K/F be a Galois extension, with Galois group $G = G(K/F)$. Then the fixed field K^G of G is F.*

PROOF. Let L denote the fixed field K^G. Then $F \subset L \subset K$ holds, and the first inclusion means that every L-automorphism of K is automatically an F-automorphism, and hence $G(K/L) \subseteq G(K/F)$. For the reverse inclusion, we note by definition of the fixed field that every element of $G = G(K/F)$ is an L-automorphism, so we have $G \subseteq G(K/L)$. Thus,

$$G(K/L) = G(K/F).$$

On the other hand, $F \subset L \subset K$ also means we have

$$[K : F] = [K : L] \cdot [L : F] .$$

Thus, $[K : L]$ divides $[K : F]$. Therefore, if we can show that $[K : F]$ divides $[K : L]$, then $[L : F] = 1$ and hence we are done.

For this, we note that we have $|G| = [K : F]$ because K/F is a Galois extension, and moreover by Theorem 5.3, we know that $|G(K/L)|$ divides $[K : L]$. But, we have already shown that $|G(K/L)| = |G(K/F)|$. Thus, $[K : L] = [K : F]$. □

REMARK 5.16. *This corollary is important because it provides a method for checking that an element of a Galois extension K/F is actually in the bottom field F. Thus, being Galois is a strong restriction on a field extension, but nevertheless there are many Galois extensions.*

Here is a converse of this corollary:

COROLLARY 5.4. *Let G be a* finite *group of automorphisms of a field K. Then K is a Galois extension of K^G, and its Galois group is G.*

PROOF. By definition of the fixed field $L = K^G$, the elements of G are automatically L-automorphisms of K. Thus, we have

$$G \subset G(K/L).$$

Moreover, K is a finite extension of L because $[K : L] = |G|$ holds by Theorem 5.2. Thus, by Theorem 5.3 $|G(K/L)|$ divides $[K : L]$; hence $|G(K/L)| \leq [K : L]$. If we put these results together, we obtain:

$$[K : L] = |G| \leq |G(K/L)| \leq [K : L] .$$

Thus, $G = G(K/L)$ and $|G(K/L)| = [K : L]$, that is, K is Galois over L and its Galois group is G. □

REMARK 5.17. *This corollary is known as Emil Artin's theorem, but it is in fact due to Dedekind (1831-1916).*

COROLLARY 5.5. *Let K/F be a Galois extension, let $g(x)$ be an irreducible polynomial in $F[x]$. If $g(x) = 0$ has one root in K, then it factors into linear factors in $K[x]$.*

PROOF. By Corollary 5.3, F is the fixed field of the Galois group $G = G(K/F)$. Let β be a root of $g(x) = 0$ in K. Then by Proposition 5.3, the irreducible polynomial for β over F is the expansion in $F[x]$ of the product:

$$(x - \beta)(x - \beta_2) \cdots (x - \beta_d) \in K[x]$$

where $\{\beta_1, ..., \beta_d\}$ is the G-orbit of β. Since $g(x)$ is the irreducible polynomial for β over F, it is equal to this product, and hence it factors into linear polynomials in $K[x]$. □

In particular, this corollary says

COROLLARY 5.6. *Every Galois extension is a splitting field of some polynomial $f(x) \in F[x]$.*

PROOF. Let K/F be a Galois extension. Then $K = F(\alpha_1, ..., \alpha_k)$. Let $f(x)$ be the product of the irreducible polynomials for α_i's over F. Then $f(x)$ splits completely in K, and hence K is a splitting field for $f(x)$. □

Exercises for Section 5.2

EXERCISE 5.4. *Suppose that K/F is a finite extension of fields. Prove that every F-monomorphism of K is an automorphism of L.*

EXERCISE 5.5. *Prove that the only automorphism of $\mathbb{R}$ is the identity map.*

3. Splitting Fields

In Proposition 4.12, we have established the existence of a splitting field of a monic polynomial $f(x)$ over a field F. In what follows, we are going to show that any two splitting fields of a given polynomial $f(x)$ are isomorphic:

Any isomorphism $\varphi\colon F \longrightarrow \tilde{F}$ of fields extends to an isomorphism $F[x] \longrightarrow \tilde{F}[x]$ between two polynomial rings which sends

$$f(x) = a_n x^n + \cdots a_0 \mapsto \tilde{f}(x) = \tilde{a}_n x^n + \cdots + \tilde{a}_0,$$

where $\tilde{a}_i = \varphi(a_i)$. Then since φ is an isomorphism, we clearly have $f(x)$ is irreducible over F if and only if $\tilde{f}(x)$ is irreducible over $\tilde{F}$.

We have a generalization of Proposition 4.9:

LEMMA 5.1. *With the above notation, let $f(x)$ be an irreducible polynomial in $F[x]$. Let α be a root of $f(x) = 0$ in an extension field K of F, and let $\tilde{\alpha}$ be a root of $\tilde{f}(x) = 0$ in an extension field $\tilde{K}$ of $\tilde{F}$. Then there exists a unique isomorphism*

$$\varphi_1\colon F(\alpha) \longrightarrow \tilde{F}(\tilde{\alpha}),$$

which sends $\alpha \mapsto \tilde{\alpha}$ and whose restriction to F is φ.

PROOF. We know that $F(\alpha)$ is isomorphic to $F[x]/(f(x))$, and $F(\tilde{\alpha})$ is isomorphic to $\tilde{F}[x]/(\tilde{f}(x))$ (Proposition 4.6. Then the isomorphism

$$F[x] \longrightarrow \tilde{F}[x]$$

we have just seen gives the isomorphism

$$F[x]/(f(x)) \longrightarrow \tilde{F}[x]/(\tilde{f}(x)).$$

Putting these isomorphisms together we obtain such an isomorphism φ_1. Since it sends the generator α of $F(\alpha)$ to the generator $\tilde{\alpha}$ of $F(\tilde{\alpha})$, it is unique. □

PROPOSITION 5.4. *Let φ: $F \longrightarrow \tilde{F}$ be an isomorphism of fields. Let $f(x)$ be a non-constant polynomial in $F[x]$, and let $\tilde{f}(x)$ be the corresponding polynomial in $\tilde{F}[x]$. Let K and $\tilde{K}$ be splitting fields for $f(x)$ and $\tilde{f}(x)$, respectively. Then there exists an isomorphism Ψ: $K \longrightarrow \tilde{K}$ whose restriction on F is φ.*

PROOF. If $f(x)$ factors into linear factors over F, then so is $\tilde{f}(x)$, and hence $K = F$ and $\tilde{K} = \tilde{F}$ in this case. Otherwise, we choose an irreducible factor $g(x)$ of $f(x)$ of degree > 1. The corresponding polynomial $\tilde{g}(x)$ will be an irreducible factor of $\tilde{f}(x)$. Let α be a root of $g(x) = 0$ in K and write $F_1 = F(\alpha)$, and let $\tilde{\alpha}$ a root of $\tilde{g}(x) = 0$ and write $\tilde{F_1} = \tilde{F}(\tilde{\alpha})$. Then by Lemma 5.1, the isomorphism φ is uniquely extended to an isomorphism

$$\varphi_1 : F_1 \longrightarrow \tilde{F_1}$$

which sends $\alpha \mapsto \tilde{\alpha}$. Since K is a splitting field for f over F, K is also a splitting field for f over the larger field F_1, and similarly, $\tilde{K}$ is the splitting field for $\tilde{f}$ over $\tilde{F_1}$. Thus, by induction on the degree of K over F, we finish the proof. □

COROLLARY 5.7. *Any two splitting fields of $f(x) \in F[x]$ over F is isomorphic.*

PROOF. We simply take $F = \tilde{F}$ and $\varphi =$ the identity map. □

REMARK 5.18. *In proving this corollary, we note that the auxiliary isomorphism φ is introduced into Proposition 5.4 in order to make the induction step of the proof work.*

With the notations of the proof of Proposition 5.4, we choose an irreducible factor $g(x)$ of $f(x)$ and one of the roots, α, of $g(x) = 0$ in K. Then any isomorphism $\Psi\colon K \longrightarrow \tilde{K}$ extending φ sends the subfield $F_1 = F(\alpha)$ of K to some subfield $\tilde{F}_1$ of $\tilde{K}$, and hence by Lemma 5.1, we have $\tilde{F}_1 = \tilde{F}(\tilde{\alpha})$, where $\tilde{\alpha} = \Psi(\alpha)$ is a root $\tilde{g}(x) = 0$ in $\tilde{K}$. Thus, the number of such isomorphisms $\leq [K : F]$. This suggests us:

LEMMA 5.2. *With the notation of Proposition 5.4, the number of isomorphisms $\Psi\colon K \longrightarrow \tilde{K}$ extending φ is equal to the degree $[K : F]$.*

PROOF. In order to extend φ to Ψ, we may start by choosing any root $\tilde{\alpha}$ of $\tilde{g}(x) = 0$ in $\tilde{K}$, and then we extend φ to $\varphi_1\colon F_1 \longrightarrow \tilde{F}_1 = \tilde{F}_1(\alpha)$ be setting $\varphi_1(\alpha) = \tilde{\alpha}$. Then since $[K : F_1] \cdot [F_1 : F] = [K : F]$, we then proceed by induction on $[K : F]$. That is, by induction hypothesis, we assume for this particular choice of φ_1 there are $[K : F_1]$ number of extensions of φ_1 to an isomorphism $\Psi\colon K \longrightarrow \tilde{K}$.

Now, since any field in this chapter is assumed to have the characteristic 0, we know by Proposition 4.15 that $\tilde{g}(x) = 0$ has the degree number of distinct roots in $\tilde{K}$ because $g(x)$ and $\tilde{g}(x)$ are irreducible over F and $\tilde{F}$, respectively. Thus, the number of choices for $\tilde{\alpha}$ for a given α is equal to the degree of g, which is $[F_1 : F]$, and hence there are $[F_1 : F]$ choices for the isomorphism φ_1. Thus, by induction hypothesis, there are in total $[K : F_1] \cdot [F_1 : F] = [K : F]$ extensions of φ to $\Psi\colon K \longrightarrow \tilde{K}$. □

THEOREM 5.4. *If K is a splitting field of a polynomial $f(x)$ over F, then K is Galois extension of F. Conversely, every Galois extension is a splitting field of some polynomial $f(x) \in F[x]$.*

PROOF. The second part is Corollary 5.6. The first part follows from Lemma 5.2. Here we take $\tilde{F} = F$, $\tilde{K} = K$, and $\varphi =$ the identity map. □

REMARK 5.19. *Since any two splitting fields K of a polynomial $f(x) \in F[x]$ are isomorphic, the Galois group $G(K/F)$ depends only on the polynomial $f(x)$ up to isomorphism. For this reason, it is often referred to as* "the" Galois group of the polynomial *over F.*

COROLLARY 5.8. *Every finite extension is contained in a Galois extension.*

PROOF. Let K be a finite field extension of a field F, and let $\{\alpha_1, ..., \alpha_n\}$ be an F-basis of K. Since any finite extension is an algebraic extension, these generators α_i are all algebraic over F. Let $f(x)$ be the product of the irreducible polynomials over F of these α_i's and let L be a splitting field of $f(x)$ over K. Then L is also a splitting field of $f(x)$ over F, and hence by Theorem 5.4 L is the Galois extension containing K. □

COROLLARY 5.9. *Let K/F be a Galois extension, and let L be an intermediate field: $F \subset L \subset K$. Then K/L is also Galois.*

PROOF. We know by Theorem 5.4 that every Galois extension K/F is a splitting field of a polynomial $f(x) \in F[x]$. Thus, K is the splitting field of $f(x)$ over F; and hence it is the splitting field of the same polynomial $f(x)$ over L. Thus again by Theorem 5.4, K is a Galois extension of L. □

The following corollary collects several criteria for an extension field to be Galois:

COROLLARY 5.10. *Let K/F be a finite field extension. Then the following are equivalent:*

- **a:** *K is a Galois extension of F.*
- **b:** *K is the splitting field of an irreducible polynomial $f(x) \in F[x]$.*
- **c:** *K is the splitting field of a polynomial $f(x) \in F[x]$.*
- **d:** *F is the fixed field for the action of the Galois group $G(K/F)$ on K.*
- **e:** *F is the fixed field for an action of a finite group of automorphisms of K.*

PROOF. We leave its proof as an exercise. □

Exercises for Section 5.3

EXERCISE 5.6. *Determine the degree over $\mathbb{Q}$ of the splitting field over $\mathbb{Q}$ of the following polynomials:*

(1) $x^2 + 1$.
(2) $x^4 - 1$.
(3) $(x^2 - 2)(x^3 - 3)$.

EXERCISE 5.7. *What is the order of the Galois group*

$$G(\mathbb{Q}(\sqrt[3]{2}, i\sqrt{3})/\mathbb{Q})?$$

EXERCISE 5.8. *What is the order of the Galois group*

$$G(\mathbb{Q}(\sqrt[3]{2}, i\sqrt{3})/\mathbb{Q}(\sqrt[3]{2}))?$$

4. The Main Theorem of Galois Theory

We have studied preliminaries for a theory which connects field theory with group theory, and we are ready to describe and prove it, the main theorem of Galois theory. the determination of the intermediate fields L, those sandwiched between F and K: $F \subset L \subset K$. The Main Theorem of Galois theory asserts that if K/F is a Galois extension, then the intermediated fields are in one-to-one correspondence with the subgroups of the Galois group.

The intermediate field corresponding to a subgroup H of $G(K/F)$ is the fixed field K^H of K, as is defined in Definition 5.4. In the other direction, if L is an intermediate field, then the Galois group $G(K/L)$ is a subgroup of $G(K/F)$:

THEOREM 5.5. *(The Main Theorem of Galois Theory) Let K be a Galois extension of a field F, and let $G = G(K/F)$, the group of F-automorphisms of K, be its Galois group.*

(1) *The assignment*

$$H \rightsquigarrow K^H$$

is a bijective map from the set of subgroups of G *to the set of intermediated fields* $F \subset L \subset K$. *Its inverse map is*

$$L \rightsquigarrow G(K/L).$$

(2) *This correspondence satisfies the following property: if we write*

$$H = G(K/L),$$

then $[K : L] = |H|$; *hence* $[L : F] = [G : H]$.

(3) *If* L, L' *are two intermediate fields and if* $H = G(K/L), H' = G(K/L')$ *are the corresponding subgroups, then* $L \subset L'$ *if and only if* $H \supset H'$.

REMARK 5.20. *We note that the bottom field* F *and the top field* K *are included among the intermediate fields of* K/F*; and hence, the trivial subgroup* $\{1\}$ *corresponds to the top field* K *and the whole group* G *corresponds to the bottom field* F. *(Corollary 5.3).*

PROOF. Let K/F be a Galois extension. We have show that the two maps

$$L \rightsquigarrow G(K/L) \qquad \text{and} \qquad H \rightsquigarrow K^H$$

are inverses to each other, that is, we are going to show that the composition of these two maps in either direction is the identity map.

Let L be an intermediate field. Then by definition of the second map, the corresponding subgroup of G is $H = G(K/L)$. By definition H acts trivially on L, so

$$L \subset K^H.$$

Now for the reverse inclusion, we note that K is a Galois extension of L by Corollary 5.9; hence $[K : L] = |H|$. Then by Theorem 5.2, $|H| = [K : K^H]$, so that $L = K^H$.

For the other direction, we start with a subgroup $H \subset G$, and let $L = K^H$. Then we have

$$H \subset G(K/L).$$

For the reverse inclusion we note that K is a Galois extension of L by Corollary 5.4 and hence $[K : L] = |G(K/L)|$. Thus, we have

$$|H| = \left[K : K^H\right] = [K : L] = |G(K/L)|,$$

and consequently $H = G(K/L)$. This shows that the two maps are inverses to each other.

For the second part, if $H = G(K/L)$, then since K is a Galois extension of $L = K^H$, we have that $[K : L] = |H|$. Thus $[L : F] = [G : H]$, because we have both $|G| = |H| \cdot [G : H]$ and $|G| = [K : F] = |H| \cdot [L : F]$.

Finally, the third part is clear from the definitions of the maps. □

A Galois group is finite, so it has finitely many subgroups. But, without the Main Theorem (Theorem 5.5), it is not obvious that there are only finitely many intermediate fields. One may think that two randomly chosen elements of a Galois extension K/F will generate two different subfields, but it is usually far from the reality. We will consider more about this at the section, Biquadratic Equations.

To have a complete picture we need one more theorem, and this requires us to introduce one more terminology:

DEFINITION 5.7. *Let L be an intermediate field: $F \subset L \subset K$, and let σ be an F-automorphism of K. The isomorphic image σL of L under σ is called a* ***conjugate subfield****.*

THEOREM 5.6. *Let K/F be a Galois extension, and let L be an intermediate field. Let $H = G(K/L)$ be the corresponding subgroup of $G = G(K/F)$. Then*

a: For any element σ of G, the subgroup of G which corresponds to the conjugate subfield σL is the conjugate subgroup $\sigma H \sigma^{-1}$. That is,

$$G(K/\sigma L) = \sigma H \sigma^{-1}.$$

b: L is a Galois extension of F if and only if H is a normal subgroup of G. If it is the case, then $G(L/F)$ is isomorphic to the quotient group G/H.

PROOF. Part a: Let τ be an element of $H = G(K/L)$ and let $\alpha' := \sigma(\alpha)$ be an element of σL. Then $\sigma\tau\sigma^{-1}(\alpha') = \sigma\tau(\alpha) = \sigma(\alpha)$; thus $\sigma\tau\sigma^{-1} \in G(K/\sigma L)$. Thus we have proved that $\sigma H\sigma^{-1} \subset G(K/\sigma L)$. Then by symmetry, or by counting elements, we see that $G(K/\sigma L) = \sigma H\sigma^{-1}$. Or,

Part b: Suppose that H is a normal subgroup of G. Thus by definition, $\sigma H\sigma^{-1} = H$ for all $\sigma \in G$. Thus, by part a, we see that

$$G(K/L) = G(K/\sigma L).$$

Since $\cdot \mapsto G(K/\cdot)$ is injective, $L = \sigma L$ for each $\sigma \in G$ (Theorem 5.5). Thus, every F-automorphism σ of K carries L to itself and hence defines an F-automorphism of L by restriction. This restriction defines a homomorphism of groups:

$$\pi\colon\ G(K/F) \longrightarrow G(L/F),$$

whose kernel is clearly $H = G(K/L)$. Thus the quotient group G/H is isomorphic to a subgroup of $G(K/F)$ (by the first isomorphism theorem for groups). Then counting degrees and orders, we see that

$$[L : F] = |G/H| \leq |G(L/F)|\,.$$

On the other hand, Theorem 5.3 in particular says that $|G(L/F)| \leq [L : F]$. Therefore, we conclude that L is a Galois extension and hence that G/H is isomorphic to $G(L/F)$.

Conversely, suppose that L/F is a Galois extension. Then L is a splitting field of some polynomial $g(x) \in F[x]$, say, $L = F(\alpha_1, ..., \alpha_d)$, where α_i are roots of $g(x) = 0$ in K. Then since any F-automorphism σ of K permutes these roots, it carries L to itself, i.e., $L = \sigma L$ for each $\sigma \in G(K/F)$. Thus, by part a, $H = \sigma H\sigma^{-1}$ for each $\sigma \in G$; that is, H is a normal subgroup of G. $\square$

Now let's consider the Galois group of a product of polynomials $f(x)g(x)$ over a field F. Let K' be a splitting field of the product $f(x)g(x)$ over F. Then K' contains a splitting field K of $f(x)$ over F and K' also contains a splitting field F' of $g(x)$. This is because, $f(x)$ and $g(x)$, respectively factors into linear factors in K'.

PROPOSITION 5.5. *Under this setting, let* $G = G(K/F), H = G(F'/F)$ *and* $I = G(K'/F)$. *Then we have*

a: *G and H are quotients of I.*

b: *I is isomorphic to a subgroup of the product group $G \times H$.*

PROOF. Since we are assuming that the bottom field F has characteristic 0, all the roots in a splitting field are distinct. Since the splitting fields K and F' are Galois extensions, and hence by part b of Theorem 5.6 they are quotients of I.

For part b, we denote the canonical surjective homomorphisms

$$\begin{aligned} I &\longrightarrow G, \\ I &\longrightarrow H \end{aligned}$$

by $\sigma \rightsquigarrow \sigma_f$ and $\sigma \rightsquigarrow \sigma_g$, respectively. Here, for instance, σ_f is, by definition, the restriction of σ on K and hence it should describe the way that σ operates on the roots of $f(x) = 0$ and σ_g should describe the way that σ operates on the roots of $g(x) = 0$. Using these two canonical maps, we define a map from I to the product group $G \times H$ by sending

$$\sigma \rightsquigarrow (\sigma_f, \sigma_g).$$

This map is clearly a homomorphism of groups. Moreover, if σ_f and σ_g are both the identity map, then their common preimage σ operates trivially on the roots of the product $f(x)g(x) = 0$, and hence σ is the identity map. Therefore, The map $I \longrightarrow G \times H$ now defined is injective and hence I is isomorphic to a subgroup of $G \times H$. □

PROPOSITION 5.6. *Let $f(x)$ be a polynomial over a field F whose Galois group G is a simple non-abelian group. Let F' be a Galois extension* *of F, with abelian Galois group. Let K' be a splitting field of f over F'. Then the Galois group $G(K'/F')$ is isomorphic to G.*

We call that a group is called **simple** if it does not have a non-trivial, proper normal subgroup.

PROOF. We first prove the proposition when $[F' : F]$ is a prime integer p. After that, we will finish our proof for a non-prime case.

Assume that $[F' : F] = p$ a prime integer and hence that $H = G(F'/F)$ is a cyclic group of order p. Then the splitting field K' will contain a splitting field of $f(x)$ over F and we call it K. Then we are in the situation of Proposition 5.5, because K' is a splitting field of $f(x)g(x)$ over F, where F' is a splitting field of some $g(x)$ over F, and hence the Galois group I of K' over F is a subgroup of $G \times H$, and it maps surjectively to G. Thus, it follows that $|G|$ divides $|I|$ because $|I| = |\ker(\phi| \cdot |im(\phi|$ where $\phi : I \longrightarrow G$ is the surjection, and $|I|$ divides $|G \times H| = |G| \cdot p$. Now if $|G| = |I|$, then the degree argument shows that $K' = K$; hence K contains the Galois extension F'. Thus H is a quotient of G by part b of Proposition 5.5. But this is impossible since G is a non-abelian simple group. Thus, the only remaining possibility is that $I = G \times H$, and by part (2) of the Main Theorem (Theorem 5.5) to the chain of fields $F \subset F' \subset K'$, we conclude that $G(K'/F') = G$.

Now we assume that $[F' : F]$ is not a prime integer. Since $G(F'/F)$ is an abelian group, there always exists a cyclic quotient group H of $G(F'/F)$ of prime order by part b of Proposition 5.5. Then this quotient determines an intermediate subfield $F_1 \subset F'$, which is a Galois extension of F and its Galois group $G(F_1/F) = H$. Let K_1 be the splitting field of $f(x)$ over F_1. Then since $[F_1 : F] = |H|$ is a prime, $G(K_1/F_1) = G$ by the previous argument. Thus, after replacing F by F_1 and K by K_1 and inductively arguing on the degree $[F' : F]$, we finish the proof. □

REMARK 5.21. *This proposition is cited in the proof of Theorem 6.5.*

Exercises for Section 5.4

EXERCISE 5.9. *Prove that the set* $(1, i, \sqrt{2}, i\sqrt{2})$ *is a basis for* $\mathbb{Q}(i, \sqrt{2})$ *over* $\mathbb{Q}$.

EXERCISE 5.10. *Determine the intermediate subfields between* $\mathbb{Q}$ *and* $\mathbb{Q}(\sqrt{2}, \sqrt{3})$.

EXERCISE 5.11. *Determine the irreducible polynomial for* $i + \sqrt{2}$ *over* $\mathbb{Q}$.

EXERCISE 5.12. *Let* $K = \mathbb{Q}(\sqrt{2}, \sqrt{3}, \sqrt{5})$.

(1) *Determine* $[K : \mathbb{Q}]$.
(2) *Prove that* K *is a Galois extension of* $\mathbb{Q}$.
(3) *Determine its Galois group.*

EXERCISE 5.13. *Let* K *be the splitting field over* $\mathbb{Q}$ *of the polynomial* $f(x) = (x^2 - 2x - 1)(x^2 - 2x - 7)$.

(1) *Determine* $G(K/\mathbb{Q})$.
(2) *Determine all intermediate subfields explicitly.*

EXERCISE 5.14. *Determine all the automorphisms of the field* $\mathbb{Q}(\sqrt[3]{2})$.

EXERCISE 5.15. *Let* $f(x)$ *be a polynomial of degree* n *over a field* F, *and let* K *be a splitting field for* $f(x)$ *over* F. *Prove that* $[K : F]$ *divides* $n!$.

EXERCISE 5.16. *Assume that we know that* π *and* e *are transcendental numbers. Let* K *be the splitting field of the polynomial* $x^3 + \pi x + 6$ *over the field* $F = \mathbb{Q}(\pi)$.

(1) *Prove that* $[K : F] = 6$.
(2) *Prove that* K *is isomorphic to the splitting field of* $x^3 + ex + 6$ *over* $\mathbb{Q}(e)$.

EXERCISE 5.17. *Let* $f(x)$ *be an irreducible cubic polynomial over* $\mathbb{Q}$ *whose Galois group is* S_3. *Determine the possible Galois group of the polynomial* $(x^3 - 1) \cdot f(x)$ *over* $\mathbb{Q}$.

5. Quadratic Equations

We recall that a field extension K/F of degree 2 over a field F is generated by any element α of $K - F$: $K = F(\alpha)$ (Corollary 4.5). Moreover, we know, for instance, by Proposition 4.11 that α is a root of an irreducible quadratic polynomial

$$f(x) = x^2 + bx + c,$$

with coefficients $b, c \in F$. Then since $\beta = -b - \alpha$ is also a root of $f(x)$, this polynomial splits into linear factors over K:

$$f(x) = (x - \alpha)(x - \beta).$$

Then since α and β are roots of the same irreducible polynomial, by Proposition 4.9, there is an isomorphism of fields:

$$\sigma\colon\ F(\alpha) \longrightarrow F(\beta),$$

which sends $\alpha \mapsto \beta$ and which is the identity map on F. Since either root α or β generate the same field K, this isomorphism σ is an F-automorphism of K.

Now, since $\alpha + \beta = b$, we have $\sigma(\alpha) + \sigma(\beta) = \sigma(\alpha + \beta) = \sigma(b) = b$, and then since $\sigma(\alpha) = \beta$, $\sigma(\beta) = b - \sigma(\alpha) = b - \beta = \alpha$. Thus, this automorphism σ switches the two roots α and β. Moreover, $\sigma^2(\alpha) = \sigma(\beta) = \alpha$; hence σ^2 is the identity map on K (Proposition 5.2). But, σ itself is not the identity map because the two roots α and β are distinct. For, if α and β are the same, then $\alpha = -b/2 \in F$, but our polynomial $f(x)$ is irreducible over F so this can not happen.

Moreover, Proposition 4.11 also says that $\delta = \alpha - \beta = \sqrt{b^2 - 4c} \in K - F$ generates K: $K = F(\delta)$, because the two roots of the irreducible polynomial $x^2 - (b^2 - 4c)$ over F are δ and $-\delta$; and hence $F(-\delta) = F(\delta) = K$. Then σ clearly interchanges the two square roots δ and $-\delta$, since $\sigma(\delta) = \sigma(\alpha) - \sigma(\beta) = \beta - \alpha = -\delta$.

In summary, whenever K is obtained by adjoining a square root δ as is explained in Proposition 5.2, we have an F-automorphism σ defined by sending $\delta \mapsto -\delta$, and hence it is not the identity map and $\sigma^2 =$ the identity map. Thus, $G(K/F) = \{1, \sigma\} = S_2$, the symmetric group of order 2, and consequently $|G(K/F)| = 2 = [K : F]$, and hence we conclude that

COROLLARY 5.11. *Every quadratic extension is a Galois extension.* □

REMARK 5.22. *We recall that the group consisting of permutations of* $\{1, 2, ..., n\}$, *i.e., bijections:* $\{1, 2, ..., n\} \longrightarrow \{1, 2, ..., n\}$ *is usually denoted by*

$$S_n.$$

EXAMPLE 5.10. *Let* $K = \mathbb{Q}(\alpha)$, *where* $\alpha = 1 + \sqrt{2}$. *Squaring both sides of* $\sqrt{2} = \alpha - 1$, *we see that the irreducible polynomial for* α *over* $\mathbb{Q}$ *is* $x^2 - 2x - 1$. *Another root of this polynomial is obviously* $\beta = 1 - \sqrt{2}$. *Then sending* $1 + \sqrt{2} \longmapsto 1 - \sqrt{2}$ *is the only* $\mathbb{Q}$*-automorphism of order* 2. *Thus, its Galois group is* S_2 *and* $\mathbb{Q}(\alpha)$ *is a Galois extension of* $\mathbb{Q}$.

REMARK 5.23. *We shall see later that a quadratic extension is a radical extension of the prime degree* 2 *(Definition 6.9).*

Exercises for Section 5.5

EXERCISE 5.18. *Let* $f(x) = ax^2 + bx + c$ *be a quadratic polynomial over* $\mathbb{Q}$. *Assume that* $a > 0$ *and derive the quadratic formula for the roots*

$$\frac{-b \pm \sqrt{D}}{2},$$

where $D = b^2 - 4ac$ *is the "discriminant" of* $f(x)$, *of the equation* $f(x) = 0$ *by completing the square.*

6. Biquadratic Equations

DEFINITION 5.8. *A field extension* K/F *is called a* ***biquadratic extension*** *if* $[K : F] = 4$ *and if* K *is generated by the roots of two irreducible quadratic polynomials over* F.

Thus, every biquadratic extension has the form

$$K = F(\alpha, \beta),$$

where $\alpha^2 = a, \beta^2 = b$ are elements of F. By Theorem 5.3 we know that $|G(K/F)|$ divides $[K : F] = 4$, and hence the order of the Galois group $G(K/F)$ is at most 4.

But, we can easily show that its order is in fact exactly 4. This is because: since $[F(\alpha, \beta) : F] = 4$ holds, we easily see that $F(\beta)$ has degree 2 over F and α is not in the field $F(\beta)$ from the formula,

$$[F(\alpha, \beta) : F] = [F(\alpha, \beta) : F(\beta)] \cdot [F(\beta) : F].$$

Thus, the polynomial $x^2 - a$ is again irreducible over $F(\beta)$. Similarly, we see that the polynomial $x^2 - b$ is irreducible over $F(\alpha)$.

This means that $K = F(\alpha, \beta) = F(\beta)(\alpha)$ is a quadratic extension of $F(\beta)$, and hence we have an $F(\beta)$-automorphism σ of K which interchanges the two roots $\pm\alpha$ of $x^2 - a$. Then since it is the identity map on $F(\beta)$, it is also the identity map on F, so σ is an F-automorphism of K. Similarly, there is an $F(\alpha)$-automorphism τ of K which interchanges the two roots $\pm\beta$ of $x^2 - b$, and hence τ is also an F-automorphism of K.

In addition to this, we note that σ^2 is the identity map on K and σ fixes β, and τ^2 is the identity map on K and τ fixes α. Moreover, we note that their composition $\sigma\tau$ is another F-automorphism of K, and one can easily check that

$$\sigma\tau = \tau\sigma,$$

and hence $(\sigma\tau)^2$ is the identity map on K.

Then since the elements α and β generated K, by Proposition 5.2 we know that an F-automorphism which fixes both α and τ is the identity map. Consequently, we have in total four F-automorphisms of K including the identity map 1:

$$G(K/F) = \{1, \sigma, \tau, \sigma\tau : \sigma^2 = 1, \tau^2 = 1, \sigma\tau = \tau\sigma\},$$

which is the so-called Klein four group.

REMARK 5.24. *We note that the Klein four group has three cyclic subgroups of order* 2*, and hence it is not a cyclic group.*

Therefore, we have shown that

COROLLARY 5.12. *Every biquadratic extension K/F is a Galois extension with $G(K/F)$ the Klein four group.* □

EXAMPLE 5.11. $\mathbb{Q}(i, \sqrt{2})$ *is a Galois extension of* $\mathbb{Q}$, *where* σ *is the complex conjugation and* τ *is the map sending* $\sqrt{2} \mapsto -\sqrt{2}$ *and fixing*

$$i = \sqrt{-1}.$$

Thus, its Galois group, the Klein four group, has in total three proper cyclic subgroups of order 2*:*

$$H_1 = \{1, \sigma\}, \ H_2 = \{1, \tau\}, \ H_3 = \{1, \sigma\tau\}.$$

Thus, by Theorem 5.5 (the Main Theorem), there are in total three proper intermediate fields, namely,

$$L_1 = \mathbb{Q}(\sqrt{2}), \ L_2 = \mathbb{Q}(i), \ L_3 = \mathbb{Q}(i\sqrt{2}).$$

EXAMPLE 5.12. *This is a continuation of the above example. Let* γ *be an element of* K, *and consider* $\mathbb{Q}(\gamma)$. *If* γ *is not contained in any of the subfields* L_i*'s,* $i = 1, 2, 3$, *then we must have* $\mathbb{Q}(i, \sqrt{2}) = \mathbb{Q}(\gamma)$, *by the Main Theorem, Theorem 5.5.*

We recall that Theorem 5.1 guarantees the existence of such a single generator for a general finite extension K/F *of a field* F *of characteristic* 0.

REMARK 5.25. *In the case of a biquadratic extension, we can determine quite easily a single generator* γ *of* K *over* F*:*

For instance, when $K = \mathbb{Q}(i, \sqrt{2})$, *we know that*

$$\{1, i, \sqrt{2}, i\sqrt{2}\}$$

is a basis for $\mathbb{Q}(i, \sqrt{2})$ *over* $\mathbb{Q}$*; and hence we may write* γ *in a unique way as a linear combination of them:*

$$\gamma = c_1 + c_2 i + c_3\sqrt{2} + c_4 i\sqrt{2}.$$

From this expression, we easily see that γ *is not contained in any of these three subfields* L_i *unless two of the coefficients* c_2, c_3, c_4 *are zero. For instance,* $i + \sqrt{2}$ *generates the whole extension* $\mathbb{Q}(i, \sqrt{2})$.

EXAMPLE 5.13. *For more about how other primitive elements for* $\mathbb{Q}(i, \sqrt{2})$ *can be determined, we return to the proof of Theorem 5.1 and Examples 5.3, 5.4, 5.5, and 5.7. In particular, Remark 5.9 is about a generalization of this line.*

Exercises for Section 5.6

EXERCISE 5.19. *Prove that the Klein four group has three cyclic subgroups of order* 2*, and hence it is not a cyclic group.*

EXERCISE 5.20. *Determine the irreducible polynomial for* $i + \sqrt{2} \in \mathbb{Q}(i, \sqrt{2})$ *over* $\mathbb{Q}$.

EXERCISE 5.21. *Determine the irreducible polynomial for* $i + 3\sqrt{2} + 5\sqrt{-3} \in \mathbb{Q}(i, \sqrt{2}, \sqrt{-3})$ *over* $\mathbb{Q}$.

CHAPTER 6

Discriminants and Galois Groups

1. Introduction

It was J.J. Sylvester (1814 - 1897) who first discovered in 1851 the discriminant of a cubic equation and he was also the first mathematician who used the name "discriminant" for such expressions of quadratic equations and those of higher order. He published important papers in 1852 and 1853, namely "*On the principle of the calculus of forms*" and "*On the theory of syzygetic relations and two rational integer functions*". In particular he used matrix theory to study higher dimensional geometry. In 1877 Sylvester accepted a chair at Johns Hopkins University and he founded in 1878 the American Journal of Mathematics, the first mathematical journal in the United States.

As we shall see soon in this chapter, the numerical invariant discriminant plays an important role in determining the Galois group of a given irreducible polynomial over $\mathbb{Q}$.

2. Transitive Subgroups of S_n

Let $f(x)$ be an irreducible polynomial over a field F, and K be its splitting field. Then we have a linear factorization of $f(x)$ in $K[x]$

$$f(x) = (x - \alpha_1)(x - \alpha_2) \cdots (x - \alpha_n).$$

But since this polynomial $f(x)$ is in $F[x]$, its expansion

$$f(x) = x^n - s_1 x^{n-1} + s_2 + \cdots + (-1)^n s_n$$

is again a polynomial over F, whose coefficients s_i in F, which are polynomials in the roots α_i, are unchanged under any permutation on the set $\{\alpha_1, ..., \alpha_n\}$, or equivalently, on the set $\{1, ..., n\}$ of indices of the α_i's, of roots of $f(x)=0$. The s_i are:

$$\begin{aligned} s_1 &= \alpha_1+\cdots+\alpha_n; \\ s_2 &= \alpha_1\alpha_2+\cdots+\alpha_{n-1}\alpha_n=\sum_{i<j}\alpha_i\alpha_j; \\ s_3 &= \sum_{i<j<k}\alpha_i\alpha_j\alpha_k; \\ &\cdots \\ s_n &= \alpha_1\alpha_2\cdots\alpha_n. \end{aligned}$$

Historically, we have a terminology:

DEFINITION 6.1. *These n polynomials s_i are called the* ***elementary symmetric functions*** *in the α_i's with integer coefficients. More generally, a polynomial $f=f(f_1,...,f_n)$ in n variables $x_1,...,x_n$ is called* ***symmetric*** *if*

$$f(x_{i_1},...,x_{i_n})=f(x_1,...,x_n)$$

for each permutation $x_{i_1},...,x_{i_n}$ of the n variables $x_1,...,x_n$.

COROLLARY 6.1. *There are in total $n!$ permutations, i.e., bijections, on the set $\{1,2,...,n\}$ of indices of the roots α_i's of $f(x)=0$, and they form a group, called the* ***symmetric group****, denoted by*

$$S_n.$$

DEFINITION 6.2. *The* ***discriminant*** *D of the polynomial*

$$f(x)=(x-\alpha_1)(x-\alpha_2)\cdots(x-\alpha_n)$$

is the product

$$D=\prod_{i<j}(\alpha_i-\alpha_j)^2=(-1)^{n(n-1)/2}\prod_{i\neq j}(\alpha_i-\alpha_j).$$

Then clearly, D is a symmetric function with integer coefficients in the α_i's.

LEMMA 6.1. *Let* $f(x) = (x-\alpha_1)(x-\alpha_2)\cdots(x-\alpha_n)$. *The discriminant* D *of* $f(x)$ *is*

$$D = (-1)^{\frac{n(n-1)}{2}} \prod_{i=1}^{n} f'(\alpha_i),$$

where $f'(x)$ *is the formal derivative of* $f(x)$.

PROOF. By the product formula for differentiation, we have

$$f'(x) = \sum_{i=1}^{n} (x-\alpha_1)\cdots(x-\alpha_{i-1})(x-\alpha_{i+1})\cdots(x-\alpha_n),$$

and hence we have

$$f'(\alpha_i) = (\alpha_i-\alpha_1)\cdots(\alpha_i-\alpha_{i-1})(\alpha_i-\alpha_{i+1})\cdots(\alpha_i-\alpha_n).$$

We then note that this is the product of differences $(\alpha_i - \alpha_j)$, for given $i, j \neq i$; consequently, we have

$$\prod_{i=1}^{n} f'(\alpha_i) = (-1)^{\frac{n(n-1)}{2}} D.$$

□

PROPOSITION 6.1. *Let* K/F *be a splitting field of a polynomial* $f(x)$ *over a field* F. *Then*

(1) *The Galois group* $G = G(K/F)$ *operates faithfully on the set*

$$\{\alpha_1, ..., \alpha_n\}$$

of roots of $f(x) = 0$.

(2) *This operation represents* G *as a subgroup of the symmetric group* S_n.

(3) *The roots form a single orbit if and only if* $f(x)$ *is irreducible over* F.

(4) *If* $f(x)$ *is irreducible polynomial of degree* n *over* F, *then* n *divides* $|G|$.

PROOF. By Theorem 5.4, we know that K/F is a Galois extension with Galois group $G = G(K/F)$. Then by the proof of Proposition 5.3 we also know that if β is an element of K whose monic irreducible polynomial over F is $g(x)$, then $g(x)$ splits completely in K, and hence the G-orbit of β is the set of all roots of $g(x) = 0$. That is, G operates *transitively* on the set of roots of an irreducible polynomial $g(x) \in F[x]$, provided that this polynomial has at least one root in K. Combination of this result with Proposition 5.2 we obtain the first three parts of the proposition.

Part 4: Let α be a root of $f(x) = 0$ in K. Then since K is a Galois extension over F and $[F(\alpha) : F] = n$, by part 2 of Theorem 5.5 we know there exists a subgroup $H = G(K/F(\alpha))$ of G such that the index $[G : H] = n$. Thus, n divides $|G|$. □

When the Galois extension K/F is realized as the splitting field of a polynomial of degree n, we view the Galois group $G = G(K/F)$ as a subgroup of the symmetric group S_n by Proposition 6.1. If the polynomial $f(x)$ is irreducible, we now know that the third part of Proposition 6.1 holds, that is, G acts **transitively** on the indices $\{1, 2, ..., n\}$. However, the same Galois extension K/F can be realized as a splitting field of many polynomials, so this representation of G as a subgroup of S_n is not unique.

EXAMPLE 6.1. *For example, Let K/F be the splitting field of an irreducible cubic equation such that $[K : F] = 6$. (We will later have such a situation in Section, Cubic Equations below. Cf. Corollary 6.3). The Galois Group $G = G(K/F)$ is represented as the whole symmetric group S_3. On the other hand, by Theorem 5.1, we know that there exists an algebraic element γ over F such that $K = F(\gamma)$. Since γ satisfies an irreducible polynomial $f(x)$ over F of degree 6, its orbit has order 6. Thus, if we consider K as the splitting field of $f(x)$ over F, then its Galois group is represented as a subgroup of S_6. This means that embedding S_3 into a bigger group does not really help us understand the actual structure of S_3, and it simply means we need some additional theory for other direction of our investigation.*

To reverse the engine, let us suppose that the Galois extension K/F with Galois group $G = G(K/F)$ is the splitting field of a polynomial

$f(x)$ and that its roots in K are $\alpha_1, ..., \alpha_n$. Then by Proposition 6.1 we can view G as a subgroup of the symmetric group S_n, and this poses the following two problems:

PROBLEM 2. (1) *Given a subgroup H of S_n, decide if $G \subset H$.*
(2) *Determine G.*

Of course, if we are able to solve the first problem for *every* subgroup H of S_n, then the second problem may also be solved because S_n is a finite group of order $n!$ In particular, when n is small all the subgroups of S_n are well-known to us, but might not be the case to Galois when he first discovered Galois Theory, because Galois was the first mathematician who introduced the concept of symmetry into mathematics.

Lagrange's approach to these problems is to look for functions of the roots which are *partially symmetric.*

DEFINITION 6.3. *Let H be a subgroup of S_n. A polynomial*

$$f(\alpha_1, ..., \alpha_n)$$

in the variable $\alpha_1, ..., \alpha_n$ is called ***partially symmetric*** *if it is left fixed by the permutations in H but not fixed by any other permutations outside H.*

The following example should convince us that the structure of S_n becomes quickly complicated even when n reaches up to 4.

EXAMPLE 6.2.

(1) $S_1 = \{1\}$.
(2) S_2 is the cyclic group C_2. We note that S_2 is transitive, i.e., its action on the set $\{1, 2\}$ has only one orbit.
(3) S_3 has three conjugate subgroups of order 2 and one subgroup of order $3 = \frac{3!}{2}$, i.e., the alternating subgroup $A_3 = \{(123)\}$ of S_3, consisting of all even permutations, and moreover A_3 is transitive:

$$(32), (31), (21), \text{ and } (231).$$

Thus, there are in total two subgroups of S_3 which are transitive; S_3 and A_3.

We also note that S_3 is a non-abelian group because we have, for instance, $(123)(23) = (13)$ but $(23)(123) = (12)$, and moreover it is a non-abelian group of least order since any group of order less than or equal to 5 is abelian.

(4) S_4 has $4! = 24$ permutations, and it is already very complicated, because there are 4 different subgroups of S_4 which are isomorphic to S_3 and 9 subgroups which are isomorphic to S_2, and one subgroup of order 12, i.e., the alternating subgroup A_4, which is transitive.

In fact, here are all the *transitive* subgroups of S_4:

(a) S_4 itself,
(b) the alternating subgroup A_4 of order 12,
(c) the dihedral group D_4 of order 8 (which is geometrically generated by a rotation and a reflection),
(d) the cyclic subgroup C_4 of order 4 (which is geometrically generated by a rotation), and finally
(e) the Klein 4 group V consisting of

$$\{(1), (12)(34), (13)(24), (14)(23)\},$$

where the notation V comes from Klein's term *Vierergruppe* or *four group*.

(5) S_4 consists of the following permutations:

$$\begin{aligned}
&(1),\\
&(12), (13), (14), (23), (24), (34),\\
&(12)(34), (13)(24), (14)(23),\\
&(123), (124), (134), (234), (132), (142), (143), (243),\\
&(1234), (1243), (2143), (2134), (1324), (1423).
\end{aligned}$$

(6) A_4 consists of the following permutations:

$$\begin{aligned}&(1),\\&(12)(34),(13)(24),(14)(23),\\&(123),(124),(134),(234),(132),(142),(143),(243).\end{aligned}$$

(7) D_4 consists of the following permutations:

$$\begin{aligned}&(1),\\&(13),(24),\\&(12)(34),(13)(24),(14)(23),\\&(1234),(1432).\end{aligned}$$

(8) For example, V acts transitively on the set $\{1,2,3,4\}$, because if we write $\sigma=(12)(34)$, $\delta=(13)(24)$, and $\tau=(14)(23)$, then we have:

$$\begin{aligned}&1\xmapsto{\sigma}2\xmapsto{\tau}3\xmapsto{\sigma}4\xmapsto{\tau}1,\\&1\xmapsto{\sigma}2\xmapsto{\delta}4\xmapsto{\sigma}3\xmapsto{\delta}1,\\&1\xmapsto{\delta}3\xmapsto{\tau}2\xmapsto{\delta}4\xmapsto{\tau}1,\\&1\xmapsto{\delta}3\xmapsto{\sigma}4\xmapsto{\delta}2\xmapsto{\sigma}1,\\&1\xmapsto{\tau}4\xmapsto{\delta}2\xmapsto{\tau}3\xmapsto{\delta}1,\\&1\xmapsto{\tau}4\xmapsto{\sigma}3\xmapsto{\tau}2\xmapsto{\sigma}1.\end{aligned}$$

That is, the whole set $\{1,2,3,4\}$ is the only one orbit under the action of V on $\{1,2,3,4\}$.

(9) Moreover, V is a normal subgroup of S_4 contained in A_4 such that $S_4/V\approx S_3$ and $A_4/V\approx C_3$. We also note that V is a normal subgroup of S_4 contained in D_4.

(10) In fact, there are, in S_4, 3 conjugate subgroups which are isomorphic to D_4 and 3 conjugate subgroups which are isomorphic to C_4, and the other subgroups are uniquely determined. In addition to this, there are some other subgroups of S_4 which are

isomorphic to the Klein 4 group, but their actions on $\{1, 2, 3, 4\}$ are not transitive.

REMARK 6.1. *Example 6.2 will play an important role in the following two Sections.*

DEFINITION 6.4. *Define*

$$\delta(\alpha) = \prod_{i<j} (\alpha_i - \alpha_j).$$

Thus, $\delta(\alpha)$ is a square root of the discriminant D of the polynomial $f(x) = (x - \alpha_1)(x - \alpha_2) \cdots (x - \alpha_n)$. In particular, we note that the effect of a permutation of the indices is to multiply $\delta = \delta(\alpha)$ by the sign of the permutation. Thus, if $K = F(\alpha_1, ..., \alpha_n)$ is the splitting field of $f(x) \in F[x]$, then $\delta := \delta(\alpha) \in K$ is fixed by each of even permutations of the roots, $\alpha_1, ..., \alpha_n$, that is, it is fixed under the action of A_n. Then the following proposition says:

PROPOSITION 6.2. *Let K/F be a Galois extension which is the splitting field of an irreducible polynomial $f(x) \in F[x]$ of degree n. Let $\alpha_1, ...\alpha_n$ be the roots of $f(x) = 0$ in K. Then $\delta = \delta(\alpha)$ is a non-zero element of K; moreover we have*

- **a:** *$\delta \in F$ if and only if the Galois group $G = G(K/F)$ is a subgroup of the alternating group A_n.*
- **b:** *In any case, the subgroup $G(K/F(\delta))$ of G is contained in A_n.*

PROOF. The case $\delta = 0$ happens only when two of the roots are equal, but this can not happen if $f(x)$ is irreducible over F, by Proposition 4.15 because we always assume that the characteristic of F is zero throughout this chapter.

Clearly, part b follows from part a when we replace F by $F(\delta)$, so we are left to prove part a:

Now, suppose that $\delta \in F$. By Proposition 6.1 we can view G as a subgroup of the symmetric group S_n. Since any odd permutation sends $\delta \mapsto -\delta$ and $\delta \neq 0$, δ is never fixed by any elements of $S_n - A_n$. On the other hand, the elements of F are fixed by each F-automorphism in $G = G(K/F)$, and hence G cannot contain an odd permutations; hence $G \subseteq A_n$. Conversely, if $\delta \notin F$, then since $K^G = F$, there must exist an element of $G = G(K/F)$ which does not fix δ. Thus, it must be an odd permutation; consequently G is not contained in A_n. $\square$

Exercises for Section 6.2

EXERCISE 6.1. *Prove that D_4 acts transitively on the set of indices* $\{1, 2, 4, 3\}$.

EXERCISE 6.2. *Prove that A_4 acts transitively on the set of indices* $\{1, 2, 4, 3\}$.

EXERCISE 6.3. *Determine all three conjugate subgroups in S_4 each of which is isomorphic to D_4.*

EXERCISE 6.4. *Determine all three conjugate subgroups in S_4 each of which is isomorphic to C_4.*

3. Cubic Equations

Let

$$f(x) = x^3 + a_2x^2 + a_1x + a_0$$

be an irreducible cubic polynomial over over a field F. By *completing the cube*, it is rewritten as

$$\begin{aligned} f(x) &= \left(x + \frac{a_2}{3}\right)^3 - a_2x^2 - \frac{a_2^2}{3}x - \frac{a_2^3}{27} + a_2x^2 + a_1x + a_0 \\ &= \left(x + \frac{a_2}{3}\right)^3 + (a_1 - \frac{a_2^2}{3})x + (a_0 - \frac{a_2^3}{27}). \end{aligned}$$

Thus, by replacing $x + \frac{a_2}{3}$ with a new variable x', we may initially assume that our irreducible polynomial $f(x)$ over F is of the form:

$$f(x) = x^3 + px + q,$$

where $p, q \in F$. Equivalently, $g(x) = f(x - \frac{a_2}{3})$ is an irreducible polynomial over F. Thus if $\alpha_1, \alpha_2, \alpha_3$ are the three roots of $f(x) = 0$, then $\beta_i = \alpha_i +$

$\frac{a_2}{3}, i = 1, 2, 3$ are clearly the three roots of $g(x) = 0$, and consequently the discriminant D of $f(x)$ and that of $g(x)$ are the same because D is defined as the product of the squares of the differences of the roots.

REMARK 6.2. *We note that our linear replacement of $x + \frac{a_2}{3}$ by a new variable x' does not change irreducibility of $f(x)$, because such a linear replacement has an obvious inverse linear replacement.*

Let K be a splitting field of $f(x)$ over F. Then it is a Galios extension of F by Theorem 5.4. Now let $\alpha_1, \alpha_2, \alpha_3$ be the three roots of $f(x) = 0$ in K, so that we have an equality between two polynomials:

$$x^3 + px + q = (x - \alpha_1)(x - \alpha_2)(x - \alpha_3).$$

Thus, expanding the right side of this equality, we obtain three relations:

$$\begin{aligned} 0 &= \alpha_1 + \alpha_2 + \alpha_3; \\ p &= \alpha_1\alpha_2 + \alpha_2\alpha_3 + \alpha_1\alpha_3; \\ q &= -\alpha_1\alpha_2\alpha_3, \end{aligned}$$

where $p, q \in F$. Then

COROLLARY 6.2. $D = -4p^3 - 27q^2 \in F$.

PROOF. Using the above three relations, we have for $i \neq j$,

$$\begin{aligned} (\alpha_i - \alpha_j)^2 &= (\alpha_i + \alpha_j)^2 - 4\alpha_i\alpha_j \\ &= (-\alpha_k)^2 - 4\{p - \alpha_k(\alpha_i + \alpha_j)\} \\ &= \alpha_k^2 - 4\{p - \alpha_k(-\alpha_k)\} \\ &= -3\alpha_k^2 - 4p, \end{aligned}$$

where $k \neq i, j$. Thus,

$$\begin{aligned}
D &= \prod_{i<j}(\alpha_i - \alpha_j)^2 \\
&= (\alpha_1 - \alpha_2)^2(\alpha_1 - \alpha_3)^2(\alpha_2 - \alpha_3)^2 \\
&= -(3\alpha_3^2 + 4p)(3\alpha_2^2 + 4p)(3\alpha_1^2 + 4p) \\
&= -\{27(\alpha_1\alpha_2\alpha_3)^2 + 36p(\alpha_1^2\alpha_2^2 + \alpha_1^2\alpha_3^2 + \alpha_2^2\alpha_3^2) \\
&\quad +48p^2(\alpha_1^2 + \alpha_2^2 + \alpha_3^2) + 64p^3\} \\
&= -\{27q^2 + 36p(\alpha_1\alpha_2 + \alpha_1\alpha_3 + \alpha_2\alpha_3)^2 - 2\alpha_1\alpha_2\alpha_3(\alpha_1 + \alpha_2 + \alpha_3) \\
&\quad +48p^2\{(\alpha_1 + \alpha_2 + \alpha_3)^2 - 2(\alpha_1\alpha_2 + \alpha_1\alpha_3 + \alpha_2\alpha_3) + 64p^3\} \\
&= -\{27q^2 + 36p^3 - 96p^3 + 64p^3\} \\
&= -4p^3 - 27q^2 \in F.
\end{aligned}$$

□

From these relations we then easily see that the third root α_3 is in the field generated by the first two roots α_1 and α_2, and hence we have a chain of fields:

$$F \subset F(\alpha_1) \subseteq K = F(\alpha_1, \alpha_2),$$

where $K = F(\alpha_1, \alpha_2, \alpha_3) = F(\alpha_1, \alpha_2)$. Since $f(x)$ is irreducible over F, the first inclusion is always proper, and consequently, we are going to have two possibilities only, namely

(6.1) either $F(\alpha_1) = K$ *or* $F(\alpha_1)$ is a proper subfield of K.

Thus, the first case occurs when the other two roots can be expressed polynomials in α_1 over F by Proposition 4.6. The second case occurs when the other two roots can not be expressed in this way.

EXAMPLE 6.3. *Let $f(x) = x^3 - 2$. Then it is irreducible over $\mathbb{Q}$ by Proposition 2.11 and it is not irreducible over $\mathbb{R}$. The three roots of this polynomial in $\mathbb{C}$ are*

$$\alpha_1 = \sqrt[3]{2}, \alpha_2 = \omega\alpha_1, \alpha_3 = \omega^2\alpha_1,$$

where $\omega = e^{2\pi i/3}$ *the third root of* 1. *Let* K *be a splitting field of* $f(x)$ *over* $\mathbb{Q}$. *Then* K *is clearly a splitting field of* $f(x)$ *over* $\mathbb{Q}(\omega)$. *We note that* $\mathbb{Q}(\omega)$ *is an extension field of degree* 2 *over* $\mathbb{Q}$. *Now, if we take* $F = \mathbb{Q}$, *then* $\mathbb{Q}(\alpha_1)$ *is properly contained in* $\mathbb{Q}(\alpha_1, \omega) = K$. *But, if we take* $F = \mathbb{Q}(\omega)$, *then* $F(\alpha_1)$ *contains* α_2, *so* $F(\alpha_1) = K$.

Since $f(x)$ has a linear factor $x - \alpha_1$ in $F(\alpha_1)$, we have

$$f(x) = (x - \alpha_1)h(x) \in F(\alpha_1)[x],$$

where $h(x)$ is a quadratic polynomial over $F(\alpha_1)$. Then from the equality in $K[x]$:

$$(x - \alpha_1)(x - \alpha_2)(x - \alpha_3) = (x - \alpha_1)h(x),$$

we see that

$$h(x) := (x - \alpha_2)(x - \alpha_3) \in K[x].$$

Therefore, $F(\alpha_1)$ is a proper subfield of K if and only if $h(x)$ is irreducible over $F(\alpha_1)$. In this case, the degree of $F(\alpha_1)(\alpha_2)$ over $F(\alpha_1)$ is 2. Therefore we have:

$$[K : F] = \begin{cases} 3 & \text{if } F(\alpha_1) = K \\ 6 & \text{otherwise.} \end{cases}$$

We then recall that since K/F is a finite extension, the order of the Galois group $G = G(K/F)$ divides the degree $[K : F]$, and hence in particular it $\leq [K : F]$ (Theorem 5.3). Moreover, since $f(x)$ is irreducible over F, Proposition 5.2 and hence Proposition 6.1 says that G is a *subgroup* of S_3 which moreover acts *transitively* on the set $\{1, 2, 3\}$ of indices.

Now, if $[K : F] = 3$, then Example 6.2 tells us that A_3 is the only such subgroup of S_3; hence G must be equal to A_3. Thus, in this case, the cyclic permutations only in A_3 of the roots and the identity permutation are the only ones which extend to F-automorphism of K. Otherwise, that is, if $[K : F] = 6$, then Example 6.2 also tells us that $S_3 = G$. Thus, in this case, any permutation of the roots is realized by an F-automorphism of K. In any case, we have seen that the roots of an irreducible cubic polynomial may have either dihedral or cyclic symmetry.

If $[K : F] = 3$, then there are no numbers > 1 dividing 3 except 3 itself, so by the main theorem (Theorem 5.5), we have no proper intermediate subfield of K which properly contains F.

Otherwise, that is, if $[K : F] = 6$, then Example 6.2 tells us that there are 3 conjugate subgroups of S_3 of order 2 and one subgroup, A_3, of order 3. Thus, by the main theorem (Theorem 5.5), we must have three obvious intermediate subfields, $F(\alpha_1), F(\alpha_2)$, and $F(\alpha_3)$ of degree $6/2 = 3$ and we must have one intermediate subfield, say, L of degree $6/3 = 2$. We note that $F(\alpha_1), F(\alpha_2)$, and $F(\alpha_3)$ are non-equal but isomorphic to each other.

Then L is a finite extension of degree 2, and hence by Proposition 4.11, we must be able to realize L by adjoining a square root δ of an element of F. Moreover, since there is only one intermediate subfield of degree 2, this square root δ must be uniquely determined. Very fortunately, the main theorem (Theorem 5.5) also suggests which square root will work:

By the main theorem (Theorem 5.5), we know that L is the fixed field of the subgroup A_3; hence all even permutations of the three roots $\alpha_1, \alpha_2, \alpha_3$ should leave the not-yet determined δ fixed, while all odd permutations do not. Thus, by Proposition 6.2, we see that

$$\delta = (\alpha_1 - \alpha_2)(\alpha_1 - \alpha_3)(\alpha_2 - \alpha_3)$$

is the required square root. Since a permutation of the three roots multiplies δ by the sign of permutation, δ is not fixed by all element of $G = G(K/F) = S_3$, so that $\delta \notin F$, but δ^2 is fixed by every permutation, and hence it belongs to $F = K^G$. by Theorem 5.4 and Corollary 5.3. We note that $\delta^2 = D$, the discriminant of $f(x)$ (Definition 6.2). We have thus proved:

COROLLARY 6.3. *The splitting field K of a cubic irreducible polynomial $f(x)$ over a field F is a Galois extension of degree either* 3 *or* 6.

(1) *If $[K : F] = 3$, then its Galois group is A_3; and hence no proper intermediate subfields of K which properly contains F.*

(2) *If* $[K:F]=6$, *then its Galois group is* S_3*; and it has three intermediate subfields* $F(\alpha_1), F(\alpha_2), F(\alpha_3)$ *of degree* 3 *and one intermediate subfield* L *of degree* 2 *generated by a square root of the discriminant* D *of* $f(x)$*:*

$$\delta = (\alpha_1-\alpha_2)(\alpha_1-\alpha_3)(\alpha_2-\alpha_3),$$

where the α_i *are the three roots in* K *of* $f(x)=0$. □

We have another corollary, which is also a corollary of Proposition 6.2:

COROLLARY 6.4. *The discriminant* D *of an irreducible* cubic *polynomial* $f(x)\in F[x]$ *is a square in* F *if and only if the degree of the splitting field is* 3.

PROOF. If D is not a square, then since $\delta^2 = D$, $\delta \notin F$; hence $[F(\delta):F]=2$. Then since $\delta \in K$, the splitting field of $f(x)$ over F, $[K:F]$ is divisible by 2, and hence by Corollary 6.3, $[K:F]=6$. Conversely, if $\delta \in F$, then every element of the Galois group $G=G(K/F)$ is required to fix δ, but odd permutations change the sign of δ; hence they can not be an element of G. Thus, G is a proper non-trivial subgroup of S_3. Then again by Corollary 6.3, $[K:F]=3$. □

EXAMPLE 6.4. *Let* $f(x)=x^3+3x+1\in\mathbb{Q}[x]$. *Since neither* $(x+1)$ *nor* $(x-1)$ *are a factor of* $f(x)$, *it is irreducible over* $\mathbb{Z}$*; hence it is irreducible over* $\mathbb{Q}$. *Let* K *be the splitting field of* $f(x)$ *over* $\mathbb{Q}$. *We want to determine* $[K:\mathbb{Q}]$.

(The "first" method) It has only one real root α *because its derivative* $f'(x)=3x^2+3>0$ *for any real* $x\in\mathbb{R}$. *Thus, other two roots are complex numbers; so* $\mathbb{Q}(\alpha)$ *is a proper subfield of the splitting field* K *of* $f(x)$ *over* $\mathbb{Q}$. *Thus, we must have* $[K:\mathbb{Q}]=6$ *by Corollary 6.3.*

(The "second" method) Or, the discriminant D *of* $f(x)$ *is:*

$$\begin{aligned} D &= -4p^3-27q^2 = -4\cdot 3^3 - 27\cdot 1 \\ &= -135, \end{aligned}$$

and hence its square root δ is not a rational number. Thus, by Corollary 6.4, $[K : F] = 6$.

Thus by Corollary 6.3, $K/\mathbb{Q}$ is a Galois extension with the Galois group $G = S_3$ and K contains a unique intermediate field $L = \mathbb{Q}(\delta)$ of degree 2, where δ is a square root of D.

EXAMPLE 6.5. *In this example, we are first going to construct an irreducible polynomial $f(x)$ over $\mathbb{Q}$ of degree 3, using a 9-th root of unity. We note that 9 is not a prime number.*

Let $\omega = e^{2\pi i/9} = \cos(2\pi/9) + i\sin(2\pi/9)$, *a complex root of* $x^9 - 1 = 0$, *and let*

$$\begin{aligned}\alpha_1 &= \omega + \omega^{-1} = \omega + \omega^8 = 2\cos(2\pi/9),\\ \alpha_2 &= \omega^2 + \omega^{-2} = 2\cos(4\pi/9),\\ \alpha_3 &= \omega^4 + \omega^{-4} = 2\cos(8\pi/9).\end{aligned}$$

They are all real numbers. In what follows, we are going to show that they are algebraic over $\mathbb{Q}$.

We obviously have a relation between α_1 and α_2:

$$\alpha_1^2 - 2 = \alpha_2.$$

Define:

$$f(x) = (x - \alpha_1)(x - \alpha_2)(x - \alpha_3).$$

Expansion of $f(x)$ gives us,

$$f(x) = x^3 - (\alpha_1 + \alpha_2 + \alpha_3)x^2 + (\alpha_1\alpha_2 + \alpha_1\alpha_3 + \alpha_2\alpha_3)x - \alpha_1\alpha_2\alpha_3.$$

Then using a Computer Algebra System(CAS) we may check that

$$\begin{aligned}\alpha_1 + \alpha_2 + \alpha_3 &= 0,\\ \alpha_1\alpha_2 + \alpha_1\alpha_3 + \alpha_2\alpha_3 &= -3,\\ \alpha_1\alpha_2\alpha_3 &= -1.\end{aligned}$$

For instance, in order to quickly derive these relations among the roots of $f(x) = 0$, we use MAPLE and the following MAPLE Commands:

———————

*a1:=2*cos(2*Pi/9); a2:=2*cos(2*Pi*2/9); a3:=2*cos(2*Pi*4/9);*
*simplify(a1+a2+a3); simplify(a1*a2+a1*a3+a2*a3);*

———————

This should establish the first two relations hold. And then

———————

simplify(a1(a1^2-2)*(-a1-(a1^2-2)));*

———————

should establish the third relation, where we use the already-verified relations: $\alpha_2 = \alpha_1^2 - 2$ *and* $\alpha_3 = -\alpha_1 - \alpha_2$.

———————

In this way, we have explicitly constructed an integral polynomial

$$f(x) = x^3 - 3x + 1.$$

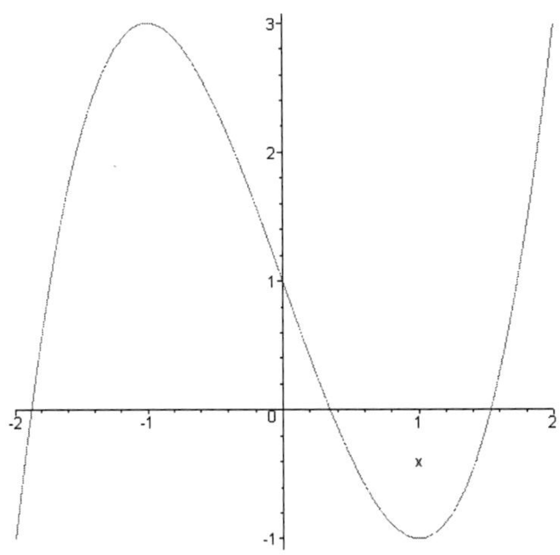

Now, since neither $(x+1)$ *nor* $(x-1)$ *is a factor of* $f(x)$, *it is irreducible over* $\mathbb{Z}$; *hence it is irreducible over* $\mathbb{Q}$. *Let* K *be the splitting field of* $f(x)$ *over* $\mathbb{Q}$. *We want to determine* $[K : \mathbb{Q}]$.

By solving $f'(x) = 0$ *for* x, *we see that* $x = -1$ *is a point where the local maximum of* $f(x)$ *occurs and* $x = 1$ *is a point where the local minimum of* $f(x)$ *occurs. Thus,* $f(x) = 0$ *has three real solutions. But, in this case, it is not easy to directly figure out these three real roots of* $f(x)$, *so that we can not try the "first" method. But, the "second" method says that*

$$\begin{aligned} D &= -4p^3 - 27q^2 \\ &= -4 \cdot (-3)^3 - 27 \cdot 1 \\ &= 81 \\ &= 9^2. \end{aligned}$$

Thus, by Corollary 6.4, we conclude that $[K : \mathbb{Q}] = 3$, *and the Galois group* $G = G(K/\mathbb{Q}) = A_3$. *Then since we have relations:*

$$\begin{aligned} \alpha_2 &= \alpha_1^2 - 2, \\ \alpha_3 &= -\alpha_1 - \alpha_2, \end{aligned}$$

the field $\mathbb{Q}(\alpha_1)$ *contains* α_2, α_3, *and consequently* $K = \mathbb{Q}(\alpha_1)$, *the splitting field of degree* 3 *of the irreducible polynomial* $f(x) = x^3 - 3x + 1$ *over* $\mathbb{Q}$. *Thus,* K *is a subfield of* $\mathbb{R}$. *Then since* 3 *is a prime number, there is no proper and non-trivial intermediate field between* $\mathbb{Q}$ *and* K.

EXAMPLE 6.6. *Another way of constructing the integral polynomial* $f(x) = x^3 - 3x + 1$ *is the following:*

Let $\omega = e^{2\pi i/9} = \cos(2\pi/9) + i\sin(2\pi/9)$, *a complex root of* $x^9 - 1 = 0$, *and let*

$$\begin{aligned} \alpha_1 &= \omega + \omega^{-1} = \omega + \omega^8 = 2\cos(2\pi/9), \\ \alpha_2 &= \omega^2 + \omega^{-2} = 2\cos(4\pi/9), \\ \alpha_3 &= \omega^4 + \omega^{-4} = 2\cos(8\pi/9). \end{aligned}$$

They are all real numbers. In what follows, we are going to show that they are algebraic over $\mathbb{Q}$.

We note that for a positive integer i *such that* $(9, i) = 1$, *i.e.,* i *being relatively prime to* 9, *each* ω^i *generates the same field* $\mathbb{Q}(\omega)$. *These generators are:*

$$\omega, \omega^2, \omega^4, \omega^5, \omega^7, \omega^8,$$

and from these elements we obtain the three real numbers $\alpha_1, \alpha_2,$*and* α_3.

Here, we use a Computer Algebra System, REDUCE.

———————

```
off echo$
off nat$
load_package groebner$
q1:=(x^9-1)/(x-1); q2:=factorize(q1);
p:=part(part(q2,1),1);
```

———————

This tells us that (the 9-th cyclotomic polynomial)

$$x^6 + x^3 + 1 = (x - \omega)(x - \omega^2)(x - \omega^4)(x - \omega^5)(x - \omega^7)(x - \omega^8) = \Phi_9(x)$$

is the irreducible polynomial for ω over $\mathbb{Q}$.

Then the following REDUCE commands compute what we are looking for:

———————

```
p1:=y-(x+x^8);
p2:=y-(x^2+x^7);
p3:=y-(x^4+x^5);
result1:=groebner({p,p1},{x,y});
result1 := {x**2 - x*y + 1,y**3 - 3*y + 1}$
result2:=groebner({p,p2},{x,y});
result2 := {x**2 + x*y**2 + x*y - 2*x + 1,y**3 - 3*y + 1}$
result3:=groebner({p,p3},{x,y});
result3 := {x**2 - x*y**2 + 2*x + 1,y**3 - 3*y + 1}$
```

———————

We note that the last polynomials of result1, result2, and result3 are the same. Thus, α_1, α_2, and α_3 are distinct roots of the same irreducible polynomial

$$f(x) = x^3 - 3x + 1 = (x - \alpha_1)(x - \alpha_2)(x - \alpha_3).$$

Thus by Proposition 6.1, the splitting field K of $f(x)$ contains all the roots α_1, α_2, and α_3; consequently we see that $K = \mathbb{Q}(\alpha_1)$ by Proposition 6.1. This is because $f(x)$ is irreducible over $\mathbb{Q}$. Thus, we conclude that

$[K:\mathbb{Q}]=3$, and hence the Galois group $G=G(K/\mathbb{Q})$ is A_3 by Corollary 6.3.

As a final task of this subsection, we are going to derive the so-called Cardano's formula for a cubic equation.

3.1. Cardano's Formula for Cubic Equations. Cubic equations

$$f(x)=x^3+a_2x^2+a_1x+a_0=0$$

were solved explicitly in terms of square roots and cube roots in the 16th century by the mathematicians Tartaglia (1500 - 1557) and Cardano (1501 - 1576).

REMARK 6.3. *More precisely, examples of cubic equations were solved by Diophantus of Alexandria (250) (200 - 280); the complete solution is due to Scipio Ferro (1515) and Niccolo Tartaglia (i.e., Nicolo of Brescia), published by Cardano in his "Ars Magna" (1545). This means that it was not Cardano who originally discovered the solution of a cubic equation.*

As we have already seen, any irreducible cubic equation

$$f(x)=x^3+a_2x^2+a_1x+a_0=0$$

is transformed to

$$f(x)=x^3+px+q=0,$$

where p,q are coefficients, i.e., already known to us.

The first step to solve this cubic equation is to put $x=u+v$ to change it to

$$u^3+v^3+q+(3uv+p)(u+v)=0. \tag{6.2}$$

Thus, if we are able to solve a system of non-linear equations for unknowns u,v,

$$\left\{\begin{array}{rcl} 3uv+p & = & 0 \\ u^3+v^3+q & = & 0 \end{array}\right.,$$

then (6.2) has a solution.

If $q = 0$, then $x^3 + px = x(x^2 + p) = 0$, so we can easily find its roots. If $q \neq 0$ and $p = 0$, then we can also easily find the three roots of

$$\left(\frac{x}{\sqrt[3]{-q}}\right)^3 - 1 = 0.$$

Thus, we may assume that $p \neq 0$ and $q \neq 0$ in solving the system of equations for u and v. Since the first equation says $v = -\frac{p}{3u}$, with this v the second equation becomes

$$3^3u^6 + 3^3qu^3 - p^3 = 0.$$

That is, we have:

$$u^6 + qu^3 - \left(\frac{p}{3}\right)^3 = 0.$$

Since this equation is a quadratic equation in the unknown u^3, one root of this equation is

$$\begin{aligned} u &= \sqrt[3]{\frac{-q + \sqrt{q^2 + 4 \cdot (p/3)^3}}{2}} \\ &= \sqrt[3]{-\frac{q}{2} + \sqrt{\left(\frac{q}{2}\right)^2 + \left(\frac{p}{3}\right)^3}}, \end{aligned}$$

and hence

$$\alpha = u + v = u - \frac{p}{3} \cdot \frac{1}{u}$$

is a root of the original equation:

$$x^3 + px + q = 0.$$

Thus, one root of the cubic equation has been solved "*by radicals*", provided that the characteristic of the bottom field is neither 2 nor 3. Then since we have

$$u^3 = \gamma^3,$$

where

$$\gamma = -\frac{q}{2} + \sqrt{\left(\frac{q}{2}\right)^2 + \left(\frac{p}{3}\right)^3},$$

we have

$$\left(\frac{u}{\gamma}\right)^3 = 1 = e^{2\pi ni} = \cos(2\pi n) + i\sin(2\pi n),$$

and hence it has three roots for $\frac{u}{\gamma}$,

$$1, \omega = e^{2\pi i/3} = \cos(2\pi/3) + i\sin(2\pi/3) = -\frac{1}{2} + \sqrt{-1}\frac{\sqrt{3}}{2}, \omega^2,$$

and hence $\left(\frac{u}{\gamma}\right)^3 = 1$ has three roots for u,

$$\gamma, \omega\gamma, \omega^2\gamma,$$

and hence the original equation

$$x^3 + px + q = 0$$

has the three roots,

$$\begin{aligned} \alpha_1 &= \gamma - \frac{p}{3} \cdot \frac{1}{\gamma}, \\ \alpha_2 &= \omega\gamma - \frac{p}{3} \cdot \frac{1}{\omega\gamma}, \\ \alpha_3 &= \omega^2\gamma - \frac{p}{3} \cdot \frac{1}{\omega^2\gamma}. \end{aligned}$$

That is, all the roots α_1, α_2, and α_3 of $x^3 + px + q = 0$ are expressed "*by radicals*".

Exercises for Section 6.3

EXERCISE 6.5. *Determine the Galois group of the following polynomials.*

(1) $x^3 - 2$.
(2) $x^3 + x + 1$.
(3) $x^3 - 3x^2 + 1$.
(4) $x^3 - x^2 - 2x + 1$.

EXERCISE 6.6. *Let $f(x)$ be an irreducible polynomial over F, and let δ be the square root of the discriminant of $f(x)$. Prove that $f(x)$ remains irreducible over the field $F(\delta)$.*

EXERCISE 6.7. *Let $f(x)$ be an irreducible cubic polynomial over $\mathbb{Q}$ which has exactly one real root., and let K be its splitting field over $\mathbb{Q}$. Prove that $[K : \mathbb{Q}] = 6$.*

EXERCISE 6.8. *When does the polynomial $x^3 + px + q$ has a multiple root?*

4. Quartic Equations

In this subsection we are going to study quartic equation, i.e., a polynomial equation of degree 4 in terms of the discriminant and partially symmetric functions.

Among quartic equations, certain types are easy to describe. We need terminologies:

DEFINITION 6.5. *Let $\alpha \in \mathbb{C}$ be a square root of a complex number and F be a subfield of $\mathbb{C}$.*

(1) A square root α is called **nested** if $\alpha = \sqrt{r + s\sqrt{t}}$ for some $r, s, t \in F$.
(2) A square root α is called **unnested** if α is a polynomial of $\sqrt{p}$, $\sqrt{q}$ for some $p, q \in F$.

Suppose that we are given a complex number α which is a nested square root, i.e.,

$$\alpha = \sqrt{r + s\sqrt{t}}$$

for some $r, s, t \in F$. Then squaring both sides, we obtain

$$\alpha^2 - r = s\sqrt{t},$$

and hence squaring both sides again, we obtain

$$\begin{aligned} 0 &= (\alpha^2 - r)^2 - (s\sqrt{t})^2 \\ &= (\alpha^2 - r - s\sqrt{t})(\alpha^2 - r + s\sqrt{t}) \\ &= \alpha^4 - 2r\alpha^2 + (r^2 - s^2 t). \end{aligned}$$

Thus, α is a root of the polynomial:

$$f(x) = (x^2 - (r + s\sqrt{t}))(x^2 - (r - s\sqrt{t})) = x^4 + bx^2 + c \in F[x],$$

where $b = -2r$ and $c = r^2 - s^2 t$. This polynomial has in total 4 roots: $\alpha, -\alpha, \alpha'$, and $-\alpha'$, where

$$\alpha' = \sqrt{r - s\sqrt{t}}$$

is one of the two square roots of $x^2 - (r - s\sqrt{t})$. Thus, α^2 and α'^2 are roots of a quadratic equation

$$q(y) = y^2 + by + c \in F[y],$$

where $y = x^2$.

If this polynomial $f(x)$ is reducible over F, then α is a root of a quadratic polynomial, and hence α is written using only one square root, so it is in fact not a nested square root at all. For example,

EXAMPLE 6.7. *Let* $\alpha = \sqrt{3 + 2\sqrt{2}}$. *Then* $b = -2r = -6$ *and* $c = 3^2 - 2^2 \cdot 2 = 1$, *so that*

$$f(x) = x^4 - 6x^2 + 1 = (x^2 + 2x - 1)(x^2 - 2x - 1),$$

and hence its four roots are

$$-1 \pm \sqrt{2}, 1 \pm \sqrt{2}.$$

Consequently, since $\alpha' = \sqrt{3 - 2\sqrt{2}} > 0$, *we conclude that*

$$\sqrt{3 + 2\sqrt{2}} = 1 + \sqrt{2}.$$

Or equivalently, we can also get the same conclusion from

$$\left(\sqrt{3 + 2\sqrt{2}}\right)^2 = \left(1 + \sqrt{2}\right)^2.$$

In order to check whether the polynomial $f(x)$ is reducible or irreducible over F, we need to check whether the polynomial $q(y)$ is reducible or irreducible over F. If $q(y)$ is irreducible then $f(x)$ can not have a root in F, and hence $f(x)$ is irreducible over F. Otherwise, namely if $q(y)$ is reducible, then $f(x)$ is going to be a product of two quadratic polynomials, and hence from the following equality with undetermined coefficients u, v, d, and e,

$$\begin{aligned} x^4 + bx^2 + c &= (x^2 + ux + v)(x^2 + dx + e) \\ &= x^4 + (u+d)x^3 + (v + e + ud)x^2 + (ue + vd)x + ve, \end{aligned}$$

we see that

$$\begin{aligned} u + d &= 0, \\ ud + v + e &= b \\ ue + vd &= 0 \\ ve &= c. \end{aligned}$$

Thus, we see that $d = -u$, and hence $e = v$ from $ue + vd = 0$, unless $u = 0$. Thus, if $u \neq 0$, we have a factorization:

$$x^4 + bx^2 + c = (x^2 + ux + v)(x^2 - ux + v).$$

If $u = 0$, then $d = 0$, and hence we have a factorization

$$x^4 + bx^2 + c = (x^2 + v)(x^2 + e).$$

Thus, we should be able to decide whether or not one of such factorizations exists, at least when $F = \mathbb{Q}$.

PROPOSITION 6.3. *Let* $\alpha = \sqrt{r + s\sqrt{t}}$ *be a nested square root for some* $r, s, t \in F$, *and assume that the polynomial*

$$f(x) = x^4 - 2rx^2 + (r^2 - s^2 t),$$

whose one root is α, *is irreducible over* F. *Then* α *can be written in terms of two unnested square roots if and only if* $r^2 - s^2 t$ *is a square in* F.

PROOF. Since $f(x)$ is assumed to be irreducible of degree 4 over F, we know that its splitting field K contains all the roots of $f(x) = 0$ by Proposition 6.1, and K is a Galois extension of F. Under this setting, we then have:

Claim 6.1. *The following statements are equivalent to each other:*

- **a:** α can be written in terms of two unnested square roots $\sqrt{p}, \sqrt{q}$.
- **b:** there exists a biquadratic extension $K = F(\sqrt{p}, \sqrt{q})$ of F which contains $\alpha = \sqrt{r + s\sqrt{t}}$.
- **c:** there exists a Galois extension K of degree 4 whose Galois group $G = G(K/F)$ is a Klein four group.

Proof of Claim. part a is clearly equivalent to part b.

Assume that part b holds. Then by Proposition 6.1, we know that K is the splitting field of the irreducible polynomial $f(x)$ over F of degree 4, which is part c.

If part c holds, then it contains the three intermediate subfields of degree 2 over F, and hence any two of these fields taken together generate K. Thus, K is a biquadratic extension of F; consequently any element of K can be written in terms of two unnested square roots, which is part a. $\square$

Suppose part a, or equivalently, part c holds. Then since the Klein 4 group V,

$$V = \{(1), (12)(34), (13)(24), (14)(23)\},$$

is the *only* transitive subgroup of the alternating group A_4 [See Example 6.2.(6)], by Proposition 6.2 we see that $\delta \in F$, where

$$\begin{aligned}
\delta^2 &= D \\
&= \prod_{i<j}(\alpha_i - \alpha_j)^2 \\
&= (4\alpha\alpha')^2(\alpha - \alpha')^4(\alpha + \alpha')^4 \\
&= 2^4 \cdot (r + s\sqrt{t})(r - s\sqrt{t}) \cdot (b^2 - 4c)^2 \\
&= 2^4 \cdot (r^2 - s^2t) \cdot \{4r^2 - 4(r^2 - s^2t)\}^2 \\
&= 2^8 s^4 t^2 (r^2 - s^2t).
\end{aligned}$$

Thus, we have seen that if part a holds, then $\delta \in F$, that is, $r^2 - s^2t$ is a square in F.

Conversely, if $r^2 - s^2t$ is a square in F, then $\delta \in F$ and hence again by Proposition 6.2, the Galois group $G = G(K/F)$ is a subgroup of A_4. But then since $f(x)$ is irreducible over F, by Proposition 6.1, the roots form a single orbit under the action of G. That is, G is transitive, and hence $G = V$, because the order of G is 4 and no other proper subgroup of A_4, except for V, acts transitively on the set of roots. (See Example 6.2). Thus, by the Claim 6.1, we finish the proof. □

EXAMPLE 6.8. (1) $\alpha = \sqrt{5+\sqrt{21}}$ *has* $r^2 - s^2t = 5^2 - 1^2 \cdot 21 = 4$, *and hence it can be written in terms of two unnested square roots, and hence it is contained in a biquadratic extension of* $\mathbb{Q}$.

(2) $\alpha = \sqrt{7+2\sqrt{5}}$ *has* $r^2 - s^2t = 7^2 - 2^2 \cdot 5 = 29$ *is not a square, and hence it can not be written in terms of two unnested square roots. Thus, it can not be contained in a biquadratic extension of* $\mathbb{Q}$.

Now, when the bottom field F is $\mathbb{Q}$, we are going to explicitly determine the two generators for the unnested expression for $\alpha = \sqrt{r+s\sqrt{t}}$ satisfying Proposition 6.3. The same method should work for any *ordered* field F when the characteristic of the bottom field F is not 2.

Since the splitting field K is a biquadratic extension of $\mathbb{Q}$, which is a Galois extension by Corollary 5.12, we know that it has three intermediate subfields of degree 2; hence each of them are generated by a square root. We know the Galois group is V,

$$V = \{(1), (12)(34), (13)(24), (14)(23)\},$$

which has three cyclic subgroups of order 2. Thus, we need to find the corresponding intermediate subfield of K. Let

$$H = \{(1), (12)(34)\},$$

and in what follows we are going to explicitly describe how to compute the corresponding subfield K^H, based on Galois correspondence between intermediate subfields and subgroups of the Galois group $G = G(K/\mathbb{Q})$.

If the four foots of $f(x)$ are listed in the following order,

$$\alpha, \alpha', -\alpha, -\alpha',$$

then under the action $\sigma = (12)(34)$, α is sent to α', and α' is sent to α. That is, the H-orbit of α is $\{\alpha, \alpha'\}$, and hence the irreducible polynomial $g(x)$ for α over K^H is,

$$g(x) = (x - \alpha)(x - \alpha') = x^2 - (\alpha + \alpha')x + \alpha\alpha'.$$

Thus, K has degree 2 over the intermediate subfield defined by

$$L = \mathbb{Q}(\alpha + \alpha', \alpha\alpha'),$$

and this subfield L is contained in K^H. Then a consideration of degrees should tell us that $L = K^H$.

Then using the hypothesis that the bottom field F is ordered, we can express $\alpha + \alpha'$ and $\alpha - \alpha'$ as a linear combination of two unnested square roots $\sqrt{u}$ and $\sqrt{v}$ for some $u, v \in F$, as the following example shows. Then finally, a consideration of degrees being equal should enable us conclude that

$$K = F(\alpha, \alpha') = F(\sqrt{u}, \sqrt{v}).$$

EXAMPLE 6.9. *Let* $\alpha = \sqrt{5 + 1 \cdot \sqrt{21}} \in \mathbb{R}$. *Its irreducible polynomial* $f(x)$ *is*

$$f(x) = x^4 + bx + c,$$

where $b = -2r = -2 \cdot 5 = 10$ *and* $c = r^2 - s^2t = 5^2 - 21 = 4$. *Then since none of the divisors* $\pm 1, \pm 2, \pm 4$ *of the constant term* 4 *is a root of* $f(x) = 0$, $f(x)$ *is irreducible over* $\mathbb{Q}$.

Now, since $\alpha' = \sqrt{5 - 1 \cdot \sqrt{21}}$, *we see that*

$$\alpha\alpha' = 2,$$

and hence

$$(\alpha + \alpha')^2 = 5 + \sqrt{21} + 5 - \sqrt{21} + 2\alpha\alpha' = 14.$$

Then since $\alpha > \alpha' > 0$, their sum is also positive; hence $\alpha + \alpha' = \sqrt{14}$, and the intermediate subfield L is $\mathbb{Q}(\sqrt{14})$.

In the same way, we obtain

$$(\alpha - \alpha')^2 = 5 + \sqrt{21} + 5 - \sqrt{21} - 2\alpha\alpha' = 6.$$

Thus, $\alpha - \alpha' = \sqrt{6}$.

Thus, we have obtained a system of two linear equations for α and α':

$$\begin{cases} \alpha + \alpha' &= \sqrt{14} \\ \alpha - \alpha' &= \sqrt{6} \end{cases}.$$

Solving this system, we find that

$$\begin{aligned} \alpha &= \frac{1}{2}(\sqrt{14} + \sqrt{6}), \\ \alpha' &= \frac{1}{2}(\sqrt{14} - \sqrt{6}), \end{aligned}$$

respectively. We note that both α and α' are elements of the newly constructed field $\mathbb{Q}(\sqrt{14}, \sqrt{6})$, and hence the splitting field $K = \mathbb{Q}(\alpha, \alpha')$ of $f(x)$ is in fact the same as $\mathbb{Q}(\sqrt{14}, \sqrt{6})$. That is,

$$\mathbb{Q}(\alpha, \alpha') = \mathbb{Q}(\sqrt{14}, \sqrt{6}).$$

Now, $y^2 - 6$ is the minimal polynomial of $\sqrt{6}$ over $\mathbb{Q}$, and since $\sqrt{14} \notin \mathbb{Q}(\sqrt{6})$ and $\sqrt{6} \notin \mathbb{Q}(\sqrt{14})$, $\left[\mathbb{Q}(\sqrt{14}, \sqrt{6}) : \mathbb{Q}\right] > 2$, and since the minimal polynomial $x^2 - 14$ of $\sqrt{14}$ over $\mathbb{Q}$ is irreducible over $\mathbb{Q}(\sqrt{6})$, we see that $x^2 - 14$ is still the minimal polynomial of $\sqrt{14}$ over $\mathbb{Q}(\sqrt{6})$, and consequently $\left[\mathbb{Q}(\sqrt{14}, \sqrt{6}) : \mathbb{Q}\right] = 4$.

Then the following lines are another way of verifying that $\mathbb{Q}(\frac{1}{2}(\sqrt{14} + \sqrt{6})) = \mathbb{Q}(\frac{1}{2}(\sqrt{14} - \sqrt{6}))$ and it is the splitting field of the following irreducible polynomial over $\mathbb{Q}$,

$$x^4 - 10x^2 + 4,$$

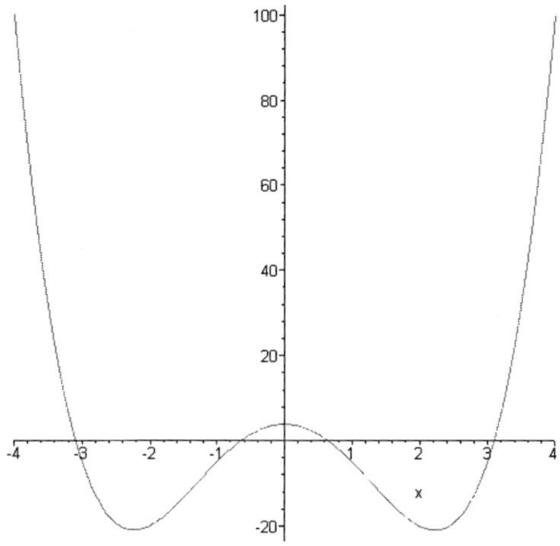

where we use a Computer Algebra System, REDUCE:

———————

```
off echo$
off nat$
load_package groebner$
p1:=x^2-14;
p2:=y^2-6;
p3:=z-(x+y)/2;
p4:=z-(x-y)/2;
result1:=groebner({p1,p2,p3},{x,y,z});
result1 := {2*x + z**3 - 12*z, 2*y - z**3 + 8*z, z**4 - 10*z**2 +
4}$
result2:=groebner({p1,p2,p4},{x,y,z});
result2 := {2*x + z**3 - 12*z, 2*y + z**3 - 8*z, z**4 - 10*z**2 +
4}$
```

———————

More precisely, the last polynomials of result1 and resut2 in the above computation are exactly the same, and hence the irreducible polynomial over $\mathbb{Q}$ *for both* $\alpha = \frac{1}{2}(\sqrt{14}+\sqrt{6})$ *and* $\alpha' = \frac{1}{2}(\sqrt{14}-\sqrt{6})$ *are the same polynomial*

$$x^4 - 10x^2 + 4.$$

In addition to this, we already know that

$$(x-\alpha)\cdot(x+\alpha)(x-\alpha')(x+\alpha') = x^4 - 10x^2 + 4,$$

which is the irreducible polynomial for $\alpha = \sqrt{5 + 1\cdot\sqrt{21}}$ *over* $\mathbb{Q}$.

Thus, we have double-checked that

$$K = \mathbb{Q}(\alpha, \alpha'),\ \mathbb{Q}(\sqrt{14}, \sqrt{6}),\ \mathbb{Q}(\frac{1}{2}(\sqrt{14}+\sqrt{6})), \mathbb{Q}(\frac{1}{2}(\sqrt{14}-\sqrt{6}))$$

are in fact the same subfield of $\mathbb{R}$.

REMARK 6.4. *In general, the splitting field* $K = F(\alpha, \alpha')$ *of the polynomial,*

$$f(x) = (x^2 - (r + s\sqrt{t}))(x^2 - (r - s\sqrt{t})) = x^4 + bx^2 + c \in F[x],$$

whose roots are $\alpha = \sqrt{r + s\sqrt{t}}$ *and* $\alpha' = \sqrt{r - s\sqrt{t}}$ *can be constructed by the sequence* $\sqrt{t}, \alpha, \alpha'$ *of three square root adjunctions, so that the degree* $[K : F]$ *divides* 8. *Such a sequence of extensions is an example of the so-called* radical extension.

In Example 6.2, we have listed all the transitive subgroups of S_4. Then when $n = 4$, Proposition 6.2 says that $\delta \in F$ if and only if $G = G(K/F)$ is a subgroup of A_4. Since our quartic polynomial $f(x)$ is assumed to be irreducible, G acts transitively on the set of roots in the splitting field K of $f(x)$ over F; hence it must be isomorphic to one of the transitive subgroups of A_4, that is, isomorphic to either A_4 or

$$V := \{(1), (12)(34), (13)(24), (14)(23)\},$$

the Klein 4 group contained in A_4.

In order to decide whether G is isomorphic to V or A_4, we need to introduce a new idea which is also somewhat based on the theory of symmetric functions and it goes as follows: First, we recall from Example

6.2 that S_4 consists of the following permutations:

$$\begin{aligned}&(1),\\&(12),(13),(14),(23),(24),(34),\\&(12)(34),(13)(24),(14)(23),\\&(123),(124),(134),(234),(132),(142),(143),(243),\\&(1234),(1243),(2143),(2134),(1324),(1423).\end{aligned}$$

Lagrange's approach to these problems is to look for functions of the roots which are *partially symmetric.* We recall Definition 6.3.

We consider the following polynomial,

$$\begin{aligned}\beta_1(\alpha)\ &:\ = \alpha_1\alpha_3 + \alpha_2\alpha_4,\\ \beta_2(\alpha)\ &:\ = \alpha_1\alpha_2 + \alpha_3\alpha_4,\\ \beta_3(\alpha)\ &:\ = \alpha_1\alpha_4 + \alpha_2\alpha_3,\end{aligned}$$

where the α_i are four distinct roots of $f(x) = 0$.

LEMMA 6.2. *$\beta_1(\alpha), \beta_2(\alpha)$, and $\beta_3(\alpha)$ are partially symmetric polynomials, left fixed by one of the three dihedral subgroups D_4 of S_4.*

PROOF. We easily note that $\beta_1(\alpha)$ is fixed by the identity permutation and those permutations in the third row of the above list of all the elements of S_4. That is, $\beta_1(\alpha)$ is left fixed by elements of V. But, they are not all of them fixing $\beta_1(\alpha)$. For, if we consider the effect of a permutation in S_4 on $\beta_1 := \beta_1(\alpha)$ one by one, we just see that

(1) $\beta_1 := \beta_1(\alpha)$ is fixed by $(1) =$ the identity permutation and

$$(13)\,,(24),(12)(34),(13)(24),(14)(23),(1234),(1432),$$

(2) β_1 is sent to $\beta_2 := \beta_2(\alpha)$,

$$\beta_2(\alpha) = \alpha_1\alpha_2 + \alpha_3\alpha_4,$$

by $(14), (23), (123), (134), (142), (243), (1243), (2134)$,

(3) β_1 is sent to $\beta_3 := \beta_3(\alpha)$,

$$\beta_3(\alpha) = \alpha_1\alpha_4 + \alpha_2\alpha_3,$$

by $(12), (34), (124), (234), (132), (143), (1324), (1423)$.

Similarly, we can check that a permutation of indices carries one $\beta_i(\alpha)$ to another $\beta_j(\alpha)$, and vice versa. Moreover we note that they are all distinct. For instance

$$\beta_1 - \beta_2 = \alpha_1\alpha_3 + \alpha_2\alpha_4 - (\alpha_1\alpha_2 + \alpha_3\alpha_4) = (\alpha_1 - \alpha_4)(\alpha_3 - \alpha_2) \neq 0,$$

because the α_i's are the distinct 4 roots of the irreducible polynomial $f(x)$ over F.

Thus, this shows that the orbit of $\beta_1 := \beta_1(\alpha)$ under the action of S_4 to the set of indices $\{1, 2, 3, 4\}$ is $\{\beta_1, \beta_2, \beta_3\}$, and hence the associated stabilizer of β_1, which is a subgroup of S_4, consists of $24/3 = 8$ elements; hence it must be one of the three dihedral subgroups D_4 of S_4. For $\beta_2(\alpha)$ and $\beta_3(\alpha)$, we apply similar arguments to finish the whole proof. □

This proof also verifies that

COROLLARY 6.5. *V is a subgroup of D_4.*

From these crucial Lagrange's observations on partially symmetric polynomials, we define

DEFINITION 6.6. *We define the* ***resolvent cubic*** *of the irreducible quartic polynomial $f(x)$ to be*

$$g(x) = (x - \beta_1(\alpha))(x - \beta_2(\alpha)(x - \beta_3(\alpha)).$$

COROLLARY 6.6. $g(x) \in F[x]$.

PROOF. We have in the above explicitly verified that this polynomial $g(x)$ is fixed by all the permutations in S_4, and hence its coefficients are symmetric functions in the roots $\alpha_1, ..., \alpha_4$ of $f(x)$. Thus, the Galois group $G(K/F)$, which is a subgroup of S_4 by Proposition 6.1, permutes the elements β_i, and hence the polynomial $g(x)$ has coefficients in the bottom field F. □

PROPOSITION 6.4. *Let $g(x)$ be the resolvent cubic of an irreducible quartic polynomial $f(x)$ over a field F, and let K be the splitting field of $f(x)$ over F. Then we have*

- **a:** *$g(x)$ has a root in F if and only if the Galois group $G = G(K/F)$ is a subgroup of one of the three dihedral groups D_4.*
- **b:** *In any case, if β is a root of $g(x)$ in K, then the Galois group $G(K/F(\beta))$ is a subgroup of a dihedral group D_4.*

PROOF. Clearly, part b follows from part a when we replace F by $F(\beta)$, so we are left to prove part a:

If $g(x)$ has a root, say β_1, in $F = K^G$, then $G = G(K/F)$ fixes β_1, and hence G is contained in the stabilizer D_4 of β_1. Conversely, if G is a subgroup of D_4, say a stabilizer of β_1, then each element of G fixes β_1; and hence by Corollary 5.3, β_1 is in $K^G = F$. □

Thus, the associated quadratic polynomial,

$$x^2 - D,$$

where D is the discriminant of the irreducible quartic polynomial $f(x)$ over F and the associated resolvent cubic of $f(x)$,

$$g(x) := (x - \beta_1(\alpha))(x - \beta_2(\alpha)(x - \beta_3(\alpha)),$$

together with Example 6.2, Proposition 6.2, and Proposition 6.4 provides a good information on the Galois group of a given irreducible quartic polynomial $f(x)$:

		D a square in F	D not a square in F
(6.3)	$g(x)$ irreducible over F	$G = A_4$	$G = S_4$
	$g(x)$ reducible over F	$G = V$	$G = D_4$ or C_4

If D is not a square in F then $G = G(K/F)$ is not a subgroup of A_4, so it cannot be either A_4 or V by Proposition Proposition 6.2. In addition to this, if $g(x)$ is moreover reducible then one of its roots is fixed by all the elements of $G = G(K/F)$, and hence G can not be S_4, because S_4 acts transitively on the roots $\{\beta_1, \beta_2, \beta_3\}$ of $g(x) = 0$ by the proof of Lemma

6.2. Thus, if D is not a square in F and $g(x)$ is reducible over F, then $G = D_4$ or C_4. If D is not a square in F and $g(x)$ is irreducible over F, then we have no other choices but $G = S_4$ from Example 6.2.

Now, if D is a square in F, then G is either A_4 or V by Proposition 6.2. If $g(x)$ is reducible over F, then at least one of its three roots is in F, and hence the number of all permutations on the set of the remaining two distinct roots of $g(x)$ is less than the number of all permutations on the set of the three roots of $g(x)$, which we are going to have when $g(x)$ is irreducible over F. Therefore, we have the first column of the table.

REMARK 6.5. *Unlike irreducible cubic equations, computation of the discriminant of an irreducible quartic equation is quite messy. But, such a computational difficulty can be easily overcome if we apply the theory of Groebner basis.*

In what follows, we are going to see the usefulness of the Theory of Groebner Basis by explicitly computing both the discriminant of an *irreducible* quartic polynomial

$$f(x) = x^4 - \sigma_1 x^3 + \sigma_2 x^2 - \sigma_3 x + \sigma_4$$

over F, where the σ_i are the elementary symmetric functions of the four distinct roots of $f(x) = 0$, and the associated resolvent cubic

$$g(x) = (x - (\alpha_1\alpha_3 + \alpha_2\alpha_4))(x - (\alpha_1\alpha_2 + \alpha_3\alpha_4))(x - (\alpha_1\alpha_4 + \alpha_2\alpha_3)).$$

of $f(x)$. For this, we need the following result:

THEOREM 6.1. *(**[CLO]**) Let $k[x_1, x_2, ..., x_n, y_1, y_2, ..., y_n]$ be the polynomial ring with a given monomial order where any monomial involving one of $x_1, ..., x_n$ is greater than all monomials in $k[y_1, y_2, ..., y_n]$. Let G be a groebner basis of the ideal in $k[x_1, ..., x_n, y_1, ..., y_n]$ generated by $\sigma_1 - y_1, \sigma_2 - y_2, ..., \sigma_n - y_n$, where the σ_i are elementary symmetric functions in the variables x_i's.. Given a polynomial $f \in k[x_1, x_2, ..., x_n]$, let $g = \bar{f}^G$ be the remainder of f on division by G. Then:*

(1) *f is symmetric if and only if $g \in k[y_1, y_2, ..., y_n]$;*

(2) *if f is symmetric, then $f = g(\sigma_1, \sigma_2, ..., \sigma_n)$ is the unique expression of f as a polynomial in the elementary symmetric polynomials $\sigma_1, \sigma_2, ..., \sigma_n$.* □

4.1. Computation of Discriminants. Using Computer Algebra System REDUCE, we first compute the groebner basis of the ideal generated by

$$\sigma_1 - y_1, \sigma_2 - y_2, \sigma_3 - y_3, \sigma_4 - y_4:$$

```
result := {x1 + x2 + x3 + x4 - y1, x2**2 + x2*x3 + x2*x4 - x2*y1
+ x3**2 + x3*x4 - x3*y1 + x4**2 - x4*y1 + y2,
    x3**3 + x3**2*x4 - x3**2*y1 + x3*x4**2 - x3*x4*y1 + x3*y2 +
x4**3 - x4**2*y1 + x4*y2 - y3,
    x4**4 - x4**3*y1 + x4**2*y2 - x4*y3 + y4}$
```

Here, $\sigma_1 =$ *x1+x2+x3+x4,*

$\sigma_2 =$ *x1*x2+x1*x3+x1*x4+x2*x3+x2*x4+x3*x4,*

$\sigma_3 =$ *x1*x2*x3+x1*x2*x4+x2*x3*x4, and* $\sigma_4 =$ *x1*x2*x3*x4*

are the four elementary symmetric functions in the four roots x1, x2, x3, x4 of $f(x) = 0$. Then we compute:

```
result3:=preduce((x1-x2)^2*(x1-x3)^2*(x1-x4)^2*(x2-x3)^2
*(x2-x4)^2*(x3-x4)^2,
result,{x1,x2,x3,x4,y1,y2,y3,y4});
result3 := - 27*y1**4*y4**2 + 18*y1**3*y2*y3*y4 - 4*y1**3*y3**3
- 4*y1**2*y2**3*y4 + y1**2*y2**2*y3**2 + 144*y1**2*y2*y4**2
- 6*y1**2*y3**2*y4 - 80*y1*y2**2*y3*y4 + 18*y1*y2*y3**3
- 192*y1*y3*y4**2 + 16*y2**4*y4 - 4*y2**3*y3**2
- 128*y2**2*y4**2 + 144*y2*y3**2*y4 - 27*y3**4 + 256*y4**3$
```

This shows that the discriminant

$$\begin{aligned} D &= (x1-x2)^2(x1-x3)^2(x1-x4)^2(x1-x5)^2 \\ &\cdot(x2-x3)^2(x2-x4)^2(x2-x5)^2 \\ &\cdot(x3-x4)^2(x3-x5)^2 \\ &\cdot(x4-x5)^2 \end{aligned}$$

is indeed a symmetric polynomial by the first part of Theorem 6.1. Moreover, by the second part of Theorem 6.1, we see that

$$\begin{aligned} (6.4)\quad D &= -27\sigma_1^4\sigma_4^2 + 18\sigma_1^3\sigma_2\sigma_3\sigma_4 - 4\sigma_1^3\sigma_3^3 - 4\sigma_1^2\sigma_2^3\sigma_4 + \sigma_1^2\sigma_2^2\sigma_3^2 \\ &+144\sigma_1^2\sigma_2\sigma_4^2 - 6\sigma_1^2\sigma_3^2\sigma_4 - 80\sigma_1\sigma_2^2\sigma_3\sigma_4 + 18\sigma_1\sigma_2\sigma_3^3 \\ &-192\sigma_1\sigma_3\sigma_4^2 + 16\sigma_2^4\sigma_4 - 4\sigma_2^3\sigma_3^2 - 128\sigma_2^2\sigma_4^2 \\ &+144\sigma_2\sigma_3^2\sigma_4 - 27\sigma_3^4 + 256\sigma_4^3 \end{aligned}$$

is the discriminant of an irreducible polynomial over F:

$$f(x) = x^4 - \sigma_1 x^3 + \sigma_2 x^2 - \sigma_3 x + \sigma_4.$$

On the other hand,

```
result4:=(x-(x1*x3+x2*x4))*(x-(x1*x2+x3*x4))
 *(x-(x1*x4+x2*x3));
result4 := x**3 - x**2*x1*x2 - x**2*x1*x3 - x**2*x1*x4
 - x**2*x2*x3 - x**2*x2*x4 - x**2*x3*x4 + x*x1**2*x2*x3
 + x*x1**2*x2*x4 + x*x1**2*x3*x4 + x*x1*x2**2*x3
 +x*x1*x2**2*x4 + x*x1*x2*x3**2 + x*x1*x2*x4**2
 + x*x1*x3**2*x4 + x*x1*x3*x4**2 + x*x2**2*x3*x4
 + x*x2*x3**2*x4 + x*x2*x3*x4**2 - x1**3*x2*x3*x4
 - x1**2*x2**2*x3**2 - x1**2*x2**2*x4**2
 - x1**2*x3**2*x4**2 - x1*x2**3*x3*x4
 - x1*x2*x3**3*x4 - x1*x2*x3*x4**3 - x2**2*x3**2*x4**2$
```

After rewriting result4 in x, we verify that each coefficient of rsult4 is a symmetric function:

```
result5:=preduce(x1*x2+x1*x3+x1*x4+x2*x3+x2*x4+x3*x4,
result,{x1,x2,x3,x4,y1,y2,y3,y4});
result5 := y2$
result6:=preduce(x1**2*x2*x3+x1**2*x2*x4+x1**2*x3*x4
+x1*x2**2*x3 +x1*x2**2*x4+x1*x2*x3**2+x1*x2*x4**2
+x1*x3**2*x4+x1*x3*x4**2 +x2**2*x3*x4+x2*x3**2*x4
+x2*x3*x4**2,result,{x1,x2,x3,x4,y1,y2,y3,y4});
result6 := y1*y3 - 4*y4$
result:=preduce(x1**3*x2*x3*x4 + x1**2*x2**2*x3**2
+ x1**2*x2**2*x4**2 + x1**2*x3**2*x4**2
+ x1*x2**3*x3*x4 + x1*x2*x3**3*x4 + x1*x2*x3*x4**3
+ x2**2*x3**2*x4**2,result,{x1,x2,x3,x4,y1,y2,y3,y4});
result := y1**2*y4 - 4*y2*y4 + y3**2$
```

Thus, the associated resolvent cubic $g(x)$ of an irreducible polynomial $f(x)$ over F is

$$g(x) = x^3 - \sigma_2 x^2 + (\sigma_1\sigma_3 - 4\sigma_4)x - (\sigma_1^2\sigma_4 - 4\sigma_2\sigma_4 + \sigma_3^2) \in F[x]. \tag{6.5}$$

In what follows, $p = \sigma_1, q = \sigma_2, r = \sigma_3$, and $s = \sigma_4$.

EXAMPLE 6.10. *The discriminant D of an irreducible polynomial,*

$$f(x) = x^4 - (-r)x + s$$

over $\mathbb{Q}$ is (with $\sigma_1 = 0, \sigma_2 = 0$)

$$\begin{aligned} D &= -27\sigma_3^4 + 256\sigma_4^3 \\ &= -27(-r)^4 + 256s^3. \end{aligned}$$

Thus if $\sigma_1 = 0$, $\sigma_2 = 0$, *then*

$$\begin{aligned} g(x) &= x^3 - 4\sigma_4 x - \sigma_3^2 \\ &= x^3 - 4sx - (-r)^2. \end{aligned}$$

EXAMPLE 6.11. *If* $p = 0, q = 0, r = 8$ *and* $s = 12$, *then by these explicitly computed results, the discriminant* D *of the irreducible polynomial*

$$f(x) = x^4 + 8x + 12$$

over $\mathbb{Q}$ *is*

$$\begin{aligned} D &= -27(-8)^4 + 256 \cdot (12)^3 \\ &= 331776 = 2^{12} \cdot 3^4, \end{aligned}$$

which is a square. Moreover, the associated resolvent cubic polynomial $g(x)$ *of* $f(x)$ *is*

$$\begin{aligned} g(x) &= x^3 - (4 \cdot 12)x - (-8)^2 \\ &= x^3 - 48x - 64, \end{aligned}$$

and we can check $g(x)$ *is irreducible over* $\mathbb{Q}$. *Therefore, the Galois group* $G = G(K/\mathbb{Q})$ *of the splitting field* K *of the irreducible polynomial* $f(x)$ *over* $\mathbb{Q}$ *is* A_4 *by the table in (6.3).*

EXAMPLE 6.12. *If* $p = 0, q = 0, r = -4$ *and* $s = 1$, *then by these explicitly computed results, the discriminant* D *of the irreducible polynomial*

$$f(x) = x^4 - 4x + 1$$

over $\mathbb{Q}$ *is*

$$D = -6656,$$

so D *is not a square in* $\mathbb{Q}$. *Moreover by (6.5), the associated resolvent cubic of* $f(x)$ *is*

$$g(x) = x^3 - 4x - 16,$$

and $g(x)$ *is irreducible over* $\mathbb{Q}$. *Thus, the Galois group* $G = G(K/\mathbb{Q})$ *of the splitting field* K *of* $f(x)$ *over* $\mathbb{Q}$ *is* S_4, *by the table in (6.3).*

EXAMPLE 6.13. *If* $f(x) = x^4 + 6x^3 + 3x^2 + 3x + 3$, *then* $f(x)$ *is irreducible over* $\mathbb{Q}$, *for instance, by Einstein, and by (6.4) we see that*

$$D = -160191,$$

and hence D *is not a square in* $\mathbb{Q}$. *Moreover by (6.5), the associated resolvent cubic of* $f(x)$ *is*

$$g(x) = x^3 - 3x^2 + 6x - 81$$

is irreducible over $\mathbb{Q}$, *because* $g(x)(\bmod 2) = x^3 + x^2 + 1$ *is irreducible over the field* $F_2 = \mathbb{Z}/2\mathbb{Z}$. *Thus, the Galois group* $G = G(K/\mathbb{Q})$ *of the splitting field* K *of* $f(x)$ *over* $\mathbb{Q}$ *is* S_4, *by the table in (6.3).*

4.2. Formula for Quartic Equations. Suppose that we are given a quartic equation:

$$x^4 + ax^3 + bx^2 + cx + d = 0.$$

By completing the quartic,

$$\begin{aligned} & x^4 + ax^3 + bx^2 + cx + d \\ = \ & (x + \frac{a}{4})^4 - ax^3 - (\frac{a^2}{8} + \frac{a^2}{4})x^2 - \frac{a^3}{16}x - \frac{a^4}{16^2} \\ & + ax^3 + bx^2 + cx + d \\ = \ & (x + \frac{a}{4})^4 + p \cdot (x + \frac{a}{4})^2 + q \cdot (x + \frac{a}{4}) + r, \end{aligned}$$

we may assume that the given quartic equation is of the form:

$$x^4 + px^2 + qx + r = 0.$$

We note that the p, q, and r are going to be real numbers if we started with the real numbers a, b, c, and d.

Then an easy look at this expression allows us, for any variable y, we have

$$x^4 + x^2y + \frac{y^2}{4} - (x^2y + \frac{y^2}{4}) + px^2 + qx + r = 0.$$

That is, by "completing the square", we have

$$(x^2 + \frac{y}{2})^2 - \{(y - p)x^2 - qx + (\frac{y^2}{4} - r)\} = 0.$$

Thus, the first part of this equation is a square of $(x^2 + \frac{y}{2}) = P$.

In particular, if we choose a special value for y for which the discriminant of the second part of the equation to be zero, i.e.,

$$q^2 - 4(y-p)(\frac{y^2}{4} - r) = 0,$$

then the second part is also a square Q^2 for some Q. For this, we solve this cubic equation for y to always have a real root, and if we, for instance, substitute this real root to the y in the second part of the equation, then the whole second part becomes of the form, Q^2 for some determined Q. Thus, for this value for y, our equation becomes

$$P^2 - Q^2 = 0,$$

which can be easily solved. Thus, our solution for x is obtained only applying the radical formulas for quadratic equations and cubic equations. Since it is quite complicated, we do not write down a complete formula for quartic equations, but this amount of explanation should convince the readers that there is such a formula.

EXAMPLE 6.14. *Using Computer Algebra System REDUCE, we can solve the four distinct roots of*

$$x^4 - 4x + 1,$$

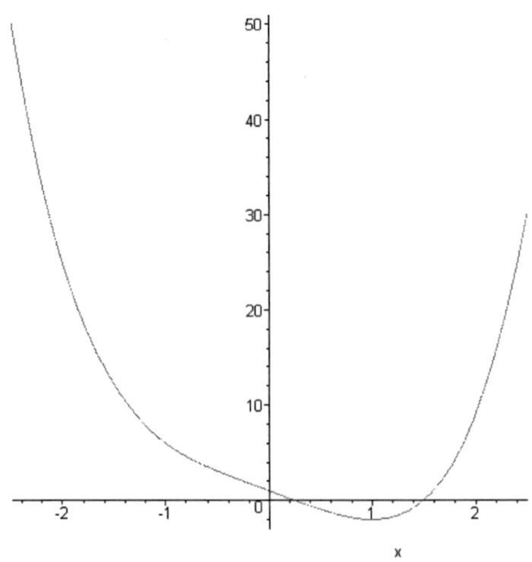

which is irreducible over $\mathbb{Q}$. For this we do the following lines within REDUCE:

———————

```
off echo$
off nat$
on fullroots;
solve(x^4-4*x+1);
Unknown: x
{x=3**((-3)/4)*(-sqrt(cosh(atanh(sqrt(26)/(3*sqrt(3)))/3))*sqrt(3)
+ sqrt( -sqrt(3)*cosh(atanh(sqrt(26)/(3*sqrt(3)))/3)
- sqrt(4*cosh(atanh(sqrt(26)/(3*sqrt(3)))/3)**2
- 3)*sqrt(3))*3**(1/4)),
x= - 3**((-3)/4)*(sqrt(cosh(atanh(sqrt(26)/(3*sqrt(3)))/3))*sqrt(3)
+ sqrt( -sqrt(3)*cosh(atanh(sqrt(26)/(3*sqrt(3)))/3)
- sqrt(4*cosh(atanh(sqrt(26)/(3*sqrt(3)))/3)**2
- 3)*sqrt(3))*3**(1/4)),
x=3**(( - 1)/4)*(sqrt(cosh(atanh(sqrt(26)/(3*sqrt(3)))/3))
+ sqrt( - cosh(atanh(sqrt(26)/(3*sqrt(3)))/3)
+ sqrt(4*cosh(atanh(sqrt(26)/(3*sqrt(3)))/3)**2 - 3))),
x=3**(( - 1)/4)*(sqrt(cosh(atanh(sqrt(26)/(3*sqrt(3)))/3))
- sqrt( - cosh(atanh(sqrt(26)/(3*sqrt(3)))/3)
+ sqrt(4*cosh(atanh(sqrt(26)/(3*sqrt(3)))/3)**2 - 3)))}$.
```

———————

Moreover, we can numerically check that the first two solutions are complex numbers, and the last two are real numbers.

Exercises for Section 6.4

EXERCISE 6.9. *Determine the Galois groups of the following polynomials over $\mathbb{Q}$.*

(1) $x^4 + 2x^2 + 4$.
(2) $x^4 - 2$.
(3) $x^4 + 2$.

(4) $x^4 + x + 1$.

EXERCISE 6.10. *Compute the discriminant of* $x^4 + 1$, *and determine its Galois group over* $\mathbb{Q}$.

EXERCISE 6.11. *Let* K *be the splitting field of the polynomial* $x^4 - 3$ *over* $\mathbb{Q}$.

(1) *Prove that* $[K : \mathbb{Q}] = 8$ *and that* K *is generated by* i *and a single root* α *of the polynomial over* $\mathbb{Q}$.
(2) *Prove that the Galois group of* $K/\mathbb{Q}$ *is dihedral, and describe the operation of the elements of* G *on the generators of* K *explicitly.*

EXERCISE 6.12. *Let* K *be the splitting field over* $\mathbb{Q}$ *of the polynomial* $x^4 - 2x^2 - 1$.

(1) *Determine the Galois group* G *of* $K/\mathbb{Q}$.
(2) *Find all intermediate subfields.*
(3) *Match them up with the subgroups of* $G = G(K/\mathbb{Q})$.

EXERCISE 6.13. *Let* K *be the splitting field of the reducible polynomial* $(x - 1)^2(x^2 + 1)$ *over* $\mathbb{Q}$. *Prove that* $\delta \in \mathbb{Q}$, *but that* $G(K/\mathbb{Q})$ *is not contained in the alternating group* A_4.

EXERCISE 6.14. *Determine whether or not the following nested radicals can be written in terms of unnested ones, and if so, find an expression.*

(1) $\sqrt{2 + \sqrt{11}}$.
(2) $\sqrt{6 + \sqrt{11}}$.
(3) $\sqrt{11 + \sqrt{6}}$.

5. Cyclotomic Extensions

DEFINITION 6.7. (1) *A* ***cyclotomic field*** *is the subfield of the field* $\mathbb{C}$ *of complex numbers which is generated over* $\mathbb{Q}$ *by* $\omega = \omega_n = e^{2\pi i/n}$.
(2) *For any subfield* F *of* $\mathbb{C}$ *the field* $F(\omega_n)$ *is called a* ***cyclotomic extension*** *of* F.

Thus, $F(\omega_n)$ is the splitting field over F of the polynomial

$$x^n - 1. \tag{6.6}$$

The roots of this polynomial are the powers of ω, the nth roots of unity $1, \omega, \omega^2, ..., \omega^{n-1}$.

In this section, we assume that n is a prime integer p which is different from 2.

We recall that

$$x^{p-1} + x^{p-2} + \cdots + 1$$

is irreducible over $\mathbb{Q}$ by Corollary 2.7, and $\omega = \omega_p = e^{2\pi i/p}$ is one of its roots. So it is the irreducible polynomial over $\mathbb{Q}$. Its roots are the distinct powers, $\omega, \omega^2, ..., \omega^{p-1}$. Thus, the extension field $\mathbb{Q}(\omega)$ over $\mathbb{Q}$, of which degree is $p-1$, is a Galois extension by Corollary 5.10 because the characteristic of $\mathbb{Q}$ is zero, and hence we have that the Galois group of the splitting field $\mathbb{Q}(\omega)$ of ω over $\mathbb{Q}$ has order $p-1$.

PROPOSITION 6.5. *Let p be an odd prime integer, and let $\omega = \omega_p = e^{2\pi i/p}$. Then*

- **a:** *The Galois group of $\mathbb{Q}(\omega)$ over $\mathbb{Q}$ is isomorphic to the multiplicative group $\mathbb{F}_p^\times$ of non-zero elements of the prime field $\mathbb{F}_p = \mathbb{Z}/p\mathbb{Z}$. It is a cyclic group of order $p-1$.*
- **b:** *For any subfield F of $\mathbb{C}$, the Galois group of $F(\omega)$ over F is a cyclic group of an order dividing $p-1$.*

PROOF. Since part a follows from part b, we first prove part b: Let $G = G(F(\omega)/F)$ be the Galois group of $F(\omega)$ over F. Let $\sigma \in G$ be an F-automorphism. Then it will carry ω to another root of the polynomial

$$x^{p-1} + x^{p-2} + \cdots + 1,$$

say, to ω^i for some positive integer i. The exponent i is determined as an integer modulo p, because ω has multiplicative order p, that is, p is the smallest positive integer such that $\omega^p = 1$. Thus, this defines a map $v \colon G \longrightarrow \mathbb{F}_p^\times$ if we set $v(\sigma) = i$. Then v is a homomorphism of groups, because if τ is another F-automorphism, sending ω to ω^j for some positive integer j, then

$$\tau\sigma(\omega) = \tau(\omega^i) = (\tau(\omega))^i = (\omega^j)^i = \omega^{ji},$$

and hence $v(\tau\sigma) = ji = v(\tau)v(\sigma)$. Thus, v is a homomorphism of groups.

Then since the identity F-automorphism fixes everything, ω is sent to itself under the identity F-automorphism. Thus, $v(1) = 1$, so that v is not an identically zero map. Moreover, since ω generates $F(\omega)$, the action of an F-automorphism on elements of $F(\omega)$ is completely determined by its action on ω. Thus, v is injective; hence G is isomorphic to its image in $\mathbb{F}_p^\times$. Since $\mathbb{F}_p^\times$ is a cyclic group by Proposition 4.18, so is every subgroup of $\mathbb{F}_p^\times$. Thus, G is cyclic of an order dividing $p-1$. In particular, if $F = \mathbb{Q}$, then $|G| = p - 1 = \left|\mathbb{F}_p^\times\right|$. $\square$

Now, we suppose that $F = \mathbb{Q}$. Since the Galois group

$$G = G(\mathbb{Q}(\omega_p)/\mathbb{Q})$$

is a cyclic group of order $p-1$, for each positive integer k dividing $p-1$, we have just one cyclic subgroup of order k. That is, if $(p-1)/k = r$ and if σ is a generator for G, then σ^r generates a subgroup of order k. Thus by the Main Theorem of Galois theory (Theorem 5.5), we have exactly one intermediate field L such that $[L : \mathbb{Q}] = r$.

PROPOSITION 6.6. *Let p be an odd prime integer, and let $\omega = \omega_p = e^{2\pi i/p}$. The subfield L of $K = \mathbb{Q}(\omega_p)$ whose degree over $\mathbb{Q}$ is $\frac{(p-1)}{2}$ is generated over $\mathbb{Q}$ by $\eta = \omega + \omega^{p-1} = 2\cos(2\pi/p)$. Moreover, $L = K \cap \mathbb{R}$.*

PROOF. Since $\omega^2 + 1 = \omega\eta$, we see that ω is a root of the quadratic equation

$$x^2 - \eta x + 1 = 0,$$

which has coefficients in $\mathbb{Q}(\eta)$. Thus, $[K : \mathbb{Q}(\eta)] \leq 2$. On the other hand, η is a real number but ω is not, so that $\mathbb{Q}(\eta)$ is a proper subfield of K. Thus it follows that $[K : \mathbb{Q}(\eta)] = 2$, that $\mathbb{Q}(\eta) = K \cap \mathbb{R}$, and that $[\mathbb{Q}(\eta) : \mathbb{Q}] = \frac{(p-1)}{2}$. $\square$

THEOREM 6.2. *Let p be an odd prime, $\omega = \omega_p = e^{2\pi i/p}$. and let L be the unique quadratic extension of $\mathbb{Q}$ contained in the cyclotomic field $\mathbb{Q}(\omega_p)$. Then*

$$L = \mathbb{Q}(\sqrt{\pm p}),$$

where the sign $\pm$ is $(-1)^{\frac{p-1}{2}}$.

PROOF. Let D be the discriminant of the polynomial

$$f(x) = x^p - 1.$$

Since $f(x) = (x-1)(x-\omega)(x-\omega^2)\cdots(x-\omega^{p-1})$ and since its derivative $f'(x) = px^{p-1}$, we have by Lemma 6.1,

$$\begin{aligned} D &= (-1)^{\frac{p(p-1)}{2}} \prod_{i=0}^{p-1} f'(\omega^i) = (-1)^{\frac{p(p-1)}{2}} \prod_{i=0}^{p-1} p\omega^{i(p-1)} \\ &= (-1)^{\frac{p(p-1)}{2}} p^p \omega^N, \end{aligned}$$

for some integer N. Then since the coefficients of $f(x)$ are rational numbers, the symmetric function D is a rational number, because we can write D as a polynomial of these rational numbers by the Theory of Invariants. But, the only power of ω which is rational is 1, and hence ω^N must be 1; consequently we have

$$\begin{aligned} D &= (-1)^{\frac{p(p-1)}{2}} p^p \\ &= (-1)^{\frac{p-1}{2}} p^p, \end{aligned}$$

since p is an odd prime integer, we have $(-1)^{\frac{p(p-1)}{2}} = ((-1)^p)^{\frac{p-1}{2}} = (-1)^{\frac{p-1}{2}}$.

Thus the square root δ of D is

$$\delta = \sqrt{\pm p^p} = \sqrt{\pm p \cdot p^{p-1}} = p^{\frac{p-1}{2}} \sqrt{\pm p} \in \mathbb{Q}(\omega)$$

by Proposition 6.2. Thus,

$$\mathbb{Q}(\delta) = \mathbb{Q}(\sqrt{\pm p})$$

is a quadratic subfield of $\mathbb{Q}(\omega)$, and since L is the only quadratic subfield, it should be L. □

A non-trivial theorem of Kronecker states that

THEOREM 6.3. *(Kronecker) Every Galois extension K of $\mathbb{Q}$ whose Galois group is abelian is contained in one of the cyclotomic fields $\mathbb{Q}(\omega_n)$.*

Such an extension is also called an **abelian extension** of $\mathbb{Q}$. We do not prove it here because it would take too long.

Now, we are going to have several examples for various primes $p = 5, 7, 17$.

EXAMPLE 6.15. *Let $p = 5$. Then $[\mathbb{Q}(\omega):\mathbb{Q}] = 5 - 1 = 4$, and hence it has a unique intermediate subfield L of degree 2 over $\mathbb{Q}$. To explicitly construct L, we note that*

$$\eta = \omega + \omega^4 = 2\cos(2\pi/5) = \frac{1}{2}(-1+\sqrt{5}) \in \mathbb{Q}(\omega).$$

Thus, the intermediate subfield L is the quadratic number field $\mathbb{Q}(\sqrt{5})$.

EXERCISE 6.15. *From the drawing of the regular pentagon with equal sides 1, verify that $\cos(\pi/5) = \frac{1}{4}(1+\sqrt{5})$.*

EXAMPLE 6.16. *Let $p = 7$. Then 6 is divisible by 3 and 2, so that we expect that it has two intermediate subfields of degree 3 and 2, respectively. In the next example we will figure out the intermediate subfield of degree 2. For the case of degree 3, we let $\omega = e^{2\pi i/7}$, and let*

$$\begin{aligned}
\alpha_1 &= \omega + \omega^6 = 2\cos(2\pi/7) \in \mathbb{Q}(\omega), \\
\alpha_2 &= \omega^2 + \omega^5 \in \mathbb{Q}(\omega), \\
\alpha_3 &= \omega^3 + \omega^4 \in \mathbb{Q}(\omega).
\end{aligned}$$

They are all real numbers. In what follows, we are going to show that they are algebraic over $\mathbb{Q}$.

We note that generators of $\mathbb{Q}(\omega)$ are:

$$\omega, \omega^2, \omega^3, \omega^4, \omega^5, \omega^6,$$

and from these elements we obtain the three real numbers α_1, α_2, and α_3.

Here, we use a Computer Algebra System, REDUCE.

———————

```
off echo$
off nat$
load_package groebner$
q1:=(x^7-1)/(x-1);
q2:=factorize(q1);
p:=part(part(q2,1),1);
```

———————

This tells us that (the 7-th cyclotomic polynomial)

$$\Phi_7(x) = x^6 + x^5 + x^4 + x^3 + x^2 + x + 1$$

is the irreducible polynomial for ω over $\mathbb{Q}$. *[This can be also checked using Eisenstein Criterion, Proposition 2.11].*

Then the following REDUCE commands compute what we are looking for:

———————

```
p1:=y-(x+x^6);
p2:=y-(x^2+x^5);
p3:=y-(x^3+x^4);
result1:=groebner({p,p1},{x,y});
result1 := {x**2 - x*y + 1,y**3 + y**2 - 2*y - 1}$
result2:=groebner({p,p2},{x,y});
result2 := {x**2 + x*y**2 + x*y - x + 1, y**3 + y**2 - 2*y - 1}$
result3:=groebner({p,p3},{x,y});
result3 := {x**2 - x*y**2 + 2*x + 1, y**3 + y**2 - 2*y - 1}$
```

———————

We note that the last polynomial of result1, result2, and result3 are the same. Thus, α_1, α_2, and α_3 are distinct roots of the same irreducible polynomial $f(x)$ *over* $\mathbb{Q}$*:*

$$f(x) = x^3 + x^2 - 2x - 1 = (x - \alpha_1)(x - \alpha_2)(x - \alpha_3).$$

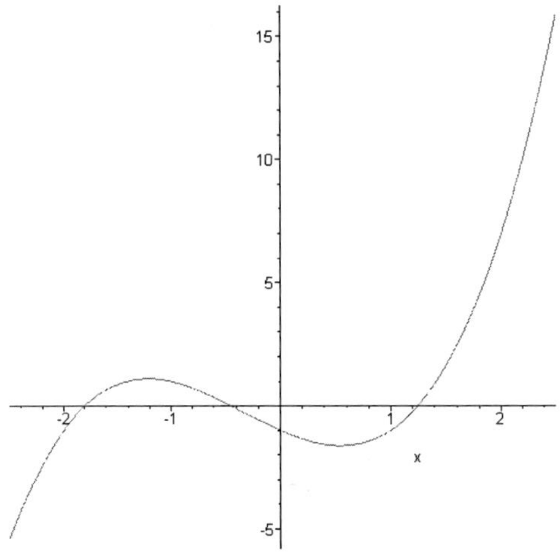

We recall that $x^3+x^2-2x-1 \pmod 2) = x^3+x^2+1$ *is irreducible over* $\mathbb{F}_2 = \mathbb{Z}/2\mathbb{Z}$ *(See Example 2.16), and hence* $f(x)$ *is irreducible over* $\mathbb{Q}$. *Thus by Proposition 6.1 and by the above computation, the splitting field* L *of* $f(x)$ *contains all the roots* α_1, α_2, *and* α_3; *consequently we see that the intermediate subfield* $L = \mathbb{Q}(\alpha_1)$ *has degree* 3.

EXAMPLE 6.17. *Let* $p = 7$, *and let* $\omega = e^{2\pi i/7}$. *We consider*

$$\begin{aligned} \alpha_1 &= \omega+\omega^2+\omega^4 \in \mathbb{Q}(\omega), \\ \alpha_2 &= \omega^3+\omega^5+\omega^6 \in \mathbb{Q}(\omega). \end{aligned}$$

Here, we use a Computer Algebra System, REDUCE.

```
off echo$
off nat$
load_package groebner$
q1:=(x^7-1)/(x-1); q2:=factorize(q1);
p:=part(part(q2,1),1);
```

This tells us that (the 7-th cyclotomic polynomial)

$$\Phi_7(x) = x^6+x^5+x^4+x^3+x^2+x+1$$

is the irreducible polynomial for ω over $\mathbb{Q}$. *[This can be also checked using Eisenstein's criterion].*

Then the following REDUCE commands compute what we are looking for:

```
————————
p1:=y-(x+x^2+x^4);
p2:=y-(x^3+x^5+x^6);
result1:=groebner({p,p1},{x,y});
result1 := {x**3 - x**2*y - x*y - x - 1,y**2 + y + 2}$
result2:=groebner({p,p2},{x,y});
result2 := {x**3 + x**2*y + x**2 + x*y - 1,y**2 + y + 2}$
————————
```

We note that the last polynomial of result1, result2, and result3 are the same. Thus, α_1, *and* α_2 are distinct roots of the same irreducible polynomial $f(x)$ *over* $\mathbb{Q}$*:*

$$f(x) = x^2 + x + 2 = (x - \alpha_1)(x - \alpha_2).$$

Thus by Proposition 6.1 and by the above computation, the splitting field M *of* $f(x)$ *contains all the roots* α_1,*and* α_2*; consequently we see that the intermediate subfield* $M = \mathbb{Q}(\alpha_1)$ *has degree* 2*. This is because* $f(x)$ *is irreducible over* $\mathbb{Q}$*. Moreover, we note that the discriminant of* $f(x)$ *is* -7*; hence* $M = \mathbb{Q}(\alpha_1) = \mathbb{Q}(\sqrt{-7})$ *is the unique intermediate subfield of degree* 2 *of* $\mathbb{Q}(\omega)$*.*

EXAMPLE 6.18. *Let* $p = 17$*, and let* $\omega = e^{2\pi i/17}$*. Since* $[\mathbb{Q}(\omega) : \mathbb{Q}] = 16 = 2^4$ *is the order of the cyclic Galois group* $C_{16} = G(\mathbb{Q}(\omega)/\mathbb{Q})$ *by Proposition 6.5,* G *has a decreasing chain of cyclic subgroups*

$$C_{16} \supset C_8 \supset C_4 \supset C_2 \supset C_1.$$

Thus, by the Main Theorem of Galois theory (Theorem 5.5), there exists a corresponding increasing chain of intermediate subfields of $\mathbb{Q}(\omega)$*:*

$$\mathbb{Q} \subset F_1 \subset F_2 \subset F_3 \subset \mathbb{Q}(\omega),$$

of degrees $1, 2, 4, 8$, *and* 16 *over* $\mathbb{Q}$. *In particular,* F_3 *of degree* $8 = (17 - 1)/2$ *is the real subfield generated by* $\eta = \omega + \omega^{16} = 2\cos(2\pi/17)$ *by Proposition 6.6. Since each extension in this chain has degree 2,* F_3 *can be reached by a succession of three square root adjunctions. This in fact proves that*

COROLLARY 6.7. $2\cos(2\pi/17)$, *and hence a regular* 17*-gon, can be constructed by ruler and compass.* □

Exercises for Section 6.5

EXERCISE 6.16. *Determine the degree of* $\omega_7 = e^{2\pi i/7}$ *over the field* $\mathbb{Q}(\omega_3)$. *Verify your result with Computer Algebra System, for instance, REDUCE.*

EXERCISE 6.17. *Let* $\omega = \omega_7$. *Determine the degree of the following elements over* $\mathbb{Q}$.

(1) $\omega + \omega^5$.
(2) $\omega^3 + \omega^4$.
(3) $\omega^3 + \omega^5 + \omega^6$.
(4) *Verify your results with Computer Algebra System, for instance, REDUCE.*

EXERCISE 6.18. *Let* $\omega = \omega_{11}$.

(1) *Prove that* $\alpha = \omega + \omega^3 + \omega^4 + \omega^5 + \omega^9$ *generates a field of degree* 2 *over* $\mathbb{Q}$, *and find its equation.*
(2) *Find an element which generates a subfield of degree* 5 *over* $\mathbb{Q}$, *and find its equation.*
(3) *Verify your results with Computer Algebra System, for instance, REDUCE.*

EXERCISE 6.19. *Let* ω *be a primitive* 13*th root of unity.*

(1) *List all intermediate subfields of* $\mathbb{Q}(\omega)$.
(2) *Determine the intermediate subfield of degree* 3 *over* $\mathbb{Q}$.

(3) *For*

$$\begin{aligned}&\omega+\omega^{12},\\&\omega+\omega^{2},\\&\omega+\omega^{5}+\omega^{8},\\&\omega^{2}+\omega^{5}+\omega^{6},\\&\omega+\omega^{5}+\omega^{8}+\omega^{12},\\&\omega+\omega^{2}+\omega^{5}+\omega^{12},\\&\omega+\omega^{3}+\omega^{4}+\omega^{9}+\omega^{10}+\omega^{12},\end{aligned}$$

determine the degree of each of them over $\mathbb{Q}$.

(4) *Verify your results with Computer Algebra System, for instance, REDUCE.*

EXERCISE 6.20. *Let* $\omega = \omega_{17}$.

(1) *Determine the succession of square roots which generate the field* $\mathbb{Q}(\omega+\omega^{16})$ *explicitly.*

(2) *Verify your results with Computer Algebra System, for instance, REDUCE.*

EXERCISE 6.21. *Prove that every quadratic extension of* $\mathbb{Q}$ *is contained in a cyclotomic extension.*

EXERCISE 6.22. *Determine the Galois group of the polynomials,* $x^8 - 1, x^{12} - 1, x^9 - 1$.

6. Kummer Extensions

Let F be a subfield of the field $\mathbb{C}$ of complex numbers which *contains* the primitive pth root of unity, $\omega_p = e^{2\pi i/p}$. We are going to study the splitting field over a field F of the polynomial of the form:

$$f(x) = x^p - a,$$

where a is a prime number. The complex roots of $f(x) = 0$ are the pth roots of a, and if we denote a particular pth root of a by α, then the distinct roots of $f(x) = 0$ are given by

$$\alpha, \omega\alpha, \omega^2\alpha, ..., \omega^{p-1}\alpha,$$

where $\omega = \omega_p = e^{2\pi i/p}$. Thus, the splitting field K of $f(x)$ over F is generated by a single root: $K = F(\alpha)$; the extension field K over F, of which degree is p, is a Galois extension by Corollary 5.10 because the characteristic of F is zero, and hence we have that the Galois group of K over F has order p.

DEFINITION 6.8. *Such a field extension is called a* ***Kummer extension.***

PROPOSITION 6.7. *Let F be a subfield of $\mathbb{C}$ which contains the pth root of unity ω_p, and let a be an element of F which is not a pth power in F, where p is a prime integer. Then the splitting field K of $f(x) = x^p - a$ has degree p over F, and its Galois group $G = G(K/F)$ is a cyclic group of order p.*

PROOF. If we can prove that $f(x)$ is irreducible over F, we are done, and it goes as follows: By hypothesis, $\alpha \notin F$. Then since the splitting field of a polynomial over F is a Galois extension and since $K^G = F$ (Corollary 5.3), there exists an F-automorphism σ of K which does not fix the element $\alpha \in K - F$. Thus, if we apply this σ to the generator α of $K = F(\alpha)$,

$$\sigma(\alpha) = \omega^\nu \alpha$$

for some non-zero ν, because the $\omega^i\alpha$'s, $i = 1, 2, ..., p-1$, are the roots of $f(x) = 0$. Now we consider

$$\sigma^2(\alpha) = \sigma(\sigma(\alpha)) = \sigma(\omega^\nu\alpha) = \omega^\nu\sigma(\alpha) = \omega^{2\nu}\alpha.$$

Thus, inductively we see that

$$\sigma^i(\sigma) = \omega^{i\nu}\alpha,$$

for $i = 1, 2, ..., p-1$. Then since ω is the pth root of unity, the smallest positive integer i such that

$$\sigma^i(\alpha) = \alpha,$$

is the prime p. This shows that the order of the Galois group $G = G(K/F)$ has order at least p. But on the other hand, since we have $K = F(\alpha)$, $[F(\alpha) : F] \leq p = \deg(f(x))$. Therefore, from these two arguments, we conclude that $[F(\alpha) : F] = p$; hence our polynomial $f(x)$ is automatically irreducible over F and the Galois group $G = G(K/F)$ is a cyclic group of order p. □

Here is the converse of Proposition 6.7:

THEOREM 6.4. *Let F be a subfield of $\mathbb{C}$ which contains the pth root of unity ω, and let K be a Galois extension of F of degree p, where p is a prime integer. Then K is obtained by adjoining a pth root to F.*

PROOF. Since the order of the Galois group $G = G(K/F)$ is a prime number p, it is a cyclic group, and hence any non-identity element σ of G will generate G. In particular, $\sigma^p = 1$.

CLAIM 6.2. *(Dedekind's Lemma) If*

$$a_0\sigma^0(u) + a_1\sigma^1(u) + \cdots + a_{p-1}\sigma^{p-1}(u) = 0,$$

where $a_i \in K$, holds for each $u \in K$, then each $a_i = 0$ for $i = 0, ..., p-1$.

PROOF OF CLAIM. (Proof by contradiction) Since $G = G(K/F)$ is a cyclic group generated by σ, the $\sigma^j(u)$ are all distinct.

Now suppose that the conclusion does not hold. Then there exist non-zero a_i's $\in K$ such that

$$a_0\sigma^0(u) + a_1\sigma(u) + \cdots + a_n\sigma^n(u) = 0 \tag{6.7}$$

holds for all $u \in K$. Among all such "dependence relations" we choose the one with minimal index n. If $n = 1$, then there exists i such that $a_i\sigma^i(u) = 0$ for each $u \in K$. Thus, if we choose ω for u, then $\sigma^i(\omega) = \omega \neq 0$, so that $a_i\sigma^i(u) = 0$ implies $a_i = 0$, which is a contradiction to to our choice of n. Thus, we may assume that $n > 1$. Since σ^i and σ^n are distinct, there exists $v \in K$ such that $\sigma^i(v) \neq \sigma^n(v)$. We apply (6.7) to the element uv for any $u \in K$ to obtain

$$a_0\sigma^0(u)\sigma^0(v) + \cdots + a_n\sigma^n(u)\sigma^n(v) = 0. \tag{6.8}$$

On the other hand, we multiply (6.7) by $\sigma^n(v)$ to obtain

$$a_0\sigma^0(u)\sigma^n(v) + \cdots + a_n\sigma^n(u)\sigma^n(v) = 0. \tag{6.9}$$

Thus, the difference of these two linear relations gives us

$$a_0(\sigma^0(v) - \sigma^n(v))\sigma^0(u) + \cdots + a_{n-1}(\sigma^{n-1}(v) - \sigma^n(v))\sigma^{n-1}(u) = 0.$$

But this dependence relation has less than n terms with coefficients $b_i = a_i(\sigma^{i-1}(v) - \sigma^n(v))$ because each $a_j \neq 0$ and $\sigma^i(v) - \sigma^n(v) \neq 0$. This is a contradiction to our choice of the "minimal" dependence relation in (6.7). This finishes the proof of Claim. □

Thus, we see that there exists a non-zero $\beta \in K$ such that

$$\alpha := \omega^0\sigma^0(\beta) + \omega\sigma(\beta) + \cdots + \omega^{p-1}\sigma^{p-1}(\beta) \neq 0.$$

We note that $\alpha \in K$, and hence $F(\alpha)$ is a subfield of K. Moreover, we have

$$\begin{aligned}
\sigma(\alpha) &= \sigma(\omega^0\sigma^0(\beta) + \omega\sigma(\beta) + \cdots + \omega^{p-1}\sigma^{p-1}(\beta)) \\
&= \omega^{-1}(\omega\omega^0\sigma(\beta) + \omega^2\sigma^2(\beta) + \cdots + \omega\omega^{p-1}\sigma^p(\beta)) \\
&= \omega^{-1}(\omega\sigma(\beta) + \omega^2\sigma^2(\beta) + \cdots + \omega^{p-1}\sigma^{p-1}(\beta) + \omega^0\sigma^0(\beta)) \\
&= \omega^{-1}\alpha = \omega^{p-1}\alpha.
\end{aligned}$$

Thus, $\alpha \neq \omega^{p-1}\alpha = \sigma(\alpha)$. That is, α is not fixed by σ, and hence $\alpha \in K - F$, because $K^G = F$. On the other hand, $\sigma(\alpha^p) = (\sigma(\alpha))^p = (\omega^{p-1}\alpha)^p = \alpha^p$ means that $\alpha^p \in F$. Thus, we are left to prove that $F(\alpha) = K$.

For this, we consider the polynomial

$$x^p - \alpha^p$$

over F. This polynomial has distinct p roots $\alpha, \omega\alpha, ..., \omega^{p-1}\alpha$, so that $F(\alpha)$ is the splitting field of this polynomial over F. Now, if can prove that this polynomial is irreducible over F, then the degree argument will establish our assertion.

To prove that $x^p - \alpha^p$ is irreducible over F, we go as follows: Since $\sigma^i(\alpha) = (\omega^{p-1})^i\alpha \in F(\alpha)$ for each $i = 0, 1, 2, ..., p-1$, its restriction to $F(\alpha)$ for a fixed i is an automorphism on $F(\alpha)$:

$$\overline{\sigma^i} : F(\alpha) \longrightarrow F(\alpha),$$

sending α to $\sigma^i(\alpha)$. But, we also have $\overline{\sigma^i}(a) = a$ for each $a \in F$. Thus, if we let

$$f(x) = x^n + a_{n-1}x^{n-1} + \cdots + a_0 \in F[x]$$

be the irreducible polynomial for α over F, then we have

$$\begin{aligned} 0 &= \overline{\sigma^i}(0) = \overline{\sigma^i}(\alpha^n + \cdots + a_1\alpha + a_0) \\ &= \overline{\sigma^i}(\alpha^n) + \overline{\sigma^i}(a_{n-1})\overline{\sigma^i}(\alpha^{n-1}) + \cdots + \overline{\sigma^i}(a_0) \\ &= (\overline{\sigma^i}(\alpha))^n + a_{n-1}(\overline{\sigma^i}(\alpha))^{n-1} + \cdots + a_0 \\ &= f(\overline{\sigma^i}(\alpha)). \end{aligned}$$

Since this holds for each $i = 0, 1, ..., p-1$, we see that the distinct p elements $(\omega^{p-1})^i\alpha = \omega^{-i}\alpha \in F(\alpha)$, for $i = 0, 1, ..., p-1$, are roots of the same irreducible polynomial $f(x) = 0$. Thus, the degree of $f(x) \geq p$. Therefore,

$$x^p - \alpha^p$$

is the irreducible polynomial for these p elements $(\omega^{p-1})^i\alpha = \omega^{-i}\alpha$ over F, and we finally conclude that $F(\alpha) = K$ from the degree argument that $[F[\alpha] : F] = p = [K : F]$ and $F(\alpha) \subseteq K$. □

EXAMPLE 6.19. *(In some sense, this example is a continuation of Example 6.6.) Let $f(x)$ be the cubic polynomial*

$$f(x) = x^3 - 3x + 1$$

over $\mathbb{Q}(\omega)$. *Let* $\{\eta_1, \eta_2, \eta_3\}$ *be its roots in the splitting field* K *of* $f(x)$ *over* $\mathbb{Q}(\omega)$. *By hypothesis, the Galois group* $G = G(K/\mathbb{Q}(\omega))$ *is of order* 3, *so that* G *has an element acting on the set of roots* $\{\eta_1, \eta_2, \eta_3\}$ *as a cyclic permutation of order* 3. *Let* $\sigma \in G = G(K/F)$ *be such an element. Then since the coefficient of* x^2 *in* $f(x)$ *is zero, we have* $\eta_1 + \eta_2 + \eta_3 = 0$, *and thus* K *is generated by* $1, \eta_1, \eta_2$ *over* $F = \mathbb{Q}(\omega)$.

In particular, we note that the ordered set $(1, \eta_1, \eta_2)$ *is a* $\mathbb{Q}(\omega)$*-basis of the vector space* K *over* $\mathbb{Q}(\omega)$. *Thus, if we consider* σ *as a linear operator on the vector space* K *over* $\mathbb{Q}(\omega)$, *the matrix of* σ *with respect to this basis is*

$$\begin{bmatrix} 1 & 0 & 0 \\ 0 & 0 & -1 \\ 0 & 1 & -1 \end{bmatrix},$$

because $\sigma(1) = 1, \sigma(\eta_1) = \eta_2, \sigma(\eta_2) = \eta_3$, *and* $\eta_3 = -\eta_1 - \eta_2$. *If we compute the eigenvalue of* σ *from the characteristic equation for this matrix, then we easily see that* $\omega = \omega_3$ *is its eigenvalue* $\neq 1$, *and*

$$\alpha = (0, 1, -\omega_3)^t = 1 \cdot \eta_1 - \omega_3 \eta_2$$

is its associated eigenvector: This is because, since $\omega_3^2 + \omega_3 + 1 = 0$ *we have* $(\omega_3 + 1)^{-1} = -\omega_3$. *Then since*

$$\sigma(\alpha) = \omega_3 \alpha$$

holds, we see that $\sigma(\alpha^3) = (\sigma(\alpha))^3 = (\omega_3\alpha)^3 = \alpha^3$, *so that* $\alpha^3 \in K^G = \mathbb{Q}(\omega)$. *In addition to this, since* $\sigma(\alpha) \neq \alpha$, *the element* α *is not in* $\mathbb{Q}(\omega)$. *Thus, using Linear Algebra a little bit, we have found an element* $\alpha \in K$ *whose third power is in* $\mathbb{Q}(\omega)$. *Then, since* $[K : \mathbb{Q}(\omega)] = 3$ *is a prime integer,* $K = \mathbb{Q}(\omega)(\alpha)$, *and* α *generates the splitting field* K *of* $x^3 - 3x + 1$ *over* $\mathbb{Q}(\omega)$.

In fact, we can compute α^3 *explicitly. That is, using the fact that* $\eta_1 = \omega_9 + \omega_9^8, \eta_2 = \omega_9^2 + \omega_9^7$, *and* $\omega_3 = \omega_9^3$ *(See Example 6.6), we find that*

$$\begin{aligned} \alpha &= \omega_9^8 - \omega_9^3(\omega_9^2 + \omega_9^7) \\ &= \omega_9^8 - \omega_9^5, \end{aligned}$$

and hence

$$\begin{aligned}\alpha^3 &= (\omega_9^8 - \omega_9^5)^3 \\ &= 3(1-\omega_3).\end{aligned}$$

Exercises for Section 6.6

EXERCISE 6.23. *Let $f(x) = x^3 + px + q$ be an irreducible polynomial over a subfield F of the complex numbers, with complex roots $\alpha = \alpha_1, \alpha_2, \alpha_3$. Let $K = F(\alpha)$.*

(1) *Express $(6\alpha^2 + 2p)^{-1}$ explicitly as a polynomial of degree 2 in α.*
(2) *Assume that $\delta = \sqrt{D}$ is in F, so that K contains the other roots of $f(x) = 0$. Express α_2 as a polynomial in $\alpha = \alpha_1$ and δ.*
(3) *Prove that $(1, \alpha, \alpha_2)$ is a basis of K, as an F-vector space.*
(4) *Let σ be the automorphism of K which permutes the three roots cyclically. Write the matrix of φ with respect to the above basis, and find its eigenvalues and eigenvectors.*
(5) *Let v be an eigenvector with eigenvalue $\omega = e^{2\pi i/3}$. Prove that $\sqrt{-3} \in F$ then $v^3 \in F$. Compute v^3 explicitly, in terms of p, q, δ, and $\sqrt{-3}$.*
(6) *Dropping the assumption that δ and $\sqrt{-3}$ are in F, express v in terms of radicals.*
(7) *Without calculation, determine the elements v' which is obtained from v by interchanging the roles of α_1 and α_2.*
(8) *Express the root a_1 in terms of radicals.*

7. Radical Extensions

The main motivation behind Galois' work was the problem of solving fifth-degree polynomial equations in one variable over a field $\mathbb{Q}$. Here is a brief history of the solution of equations by radicals. The solution of the quadratic was known to the Hindus and in its geometric form to the Greeks. The cubic and quartic were solved by the Renaissance Italian mathematicians Scipio del Ferro (1515) and Ferrari (1545) . However, not until the19th century did Abel (1802 - 1829) and Galois (1811 - 1832) prove the impossibility of solving all polynomial equations of degree $n \geq 5$.

A short time earlier than Galois, Abel proved that the quintic equation

$$x^5 + a_4x^4 + a_3x^3 + a_2x^2 + a_1x + a_0 \tag{6.10}$$

with variable rational coefficients a_i could not be solved in terms of radicals, but it remained to find an explicit polynomial in one variable with rational coefficients which could not be solved by radicals. (Our answer to this question is given in Lemma 6.4 below). Because the problem was over 200 years old before Abel and Galois, interest in it continued. It also turned out that Galois' original ideal have turned out to be much more important than the questions which motivated them.

History also tells us that Johann Carl Friedrich Gauss's (1777-1855) doctoral dissertation was a discussion of the fundamental theorem of algebra in 1799 and Evariste Galois lived during $1811-1832$, so we now easily assume that he knew all about the mathematical usefulness of complex numbers. Thus, all through out this Chapter, we have always assumed that the characteristic of the bottom field in a field extension is 0 (i.e., ∞). On the other hand, the origins and history of finite fields can be traced back to the 17th and 18th centuries, but there, it is known that these fields played only a minor role in the mathematics of the day.

We start with a

DEFINITION 6.9. *Let F be a subfield of the field $\mathbb{C}$ of complex numbers. We say that a complex number α is expressible by* **radicals** *over F if there exists an increasing chain of subfields*

$$F = F_0 \subset F_1 \subset \cdots \subset F_r$$

of $\mathbb{C}$ such that

(1) *$\alpha \in F_r$, and*
(2) *for each $j = 1, ..., r$, F_j is generated over F_{j-1} by a radical β_j. That is, $F_j = F_{j-1}(\beta_j)$ where $\beta_j^{n_j} \in F_{j-1}$ for some positive integer n_j.*

PROPOSITION 6.8. *Let $f(x)$ be an irreducible polynomial over a field F. If one root of $f(x) = 0$ in K can be expressed by radicals, the so can any other root.*

PROOF. Assume that one root α can be expressed by radicals, say using the tower

$$F = F_0 \subset F_1 \subset \cdots \subset F_r.$$

Let L be a splitting field of a polynomial of the form $f(x)g(x)$ over F for some polynomial $g(x)$ over F. Then L contains F_r and L is automatically the splitting field of $f(x)g(x)$ over $F(\alpha)$. Let α' be another root of $f(x) = 0$ in another field K', and let L' be a splitting field of $f(x)g(x)$ over $F(\alpha')$. By Proposition 5.4 the isomorphism

$$F(\alpha) \longrightarrow F(\alpha')$$

is extended to an isomorphism φ: $L \longrightarrow L'$. Then the increasing chain of subfields

$$F = \varphi(F_0) \subset \cdots \subset \varphi(F_r)$$

shows that α' is expressible by radicals. $\square$

REMARK 6.6. *The nth roots of unity $\omega_n = e^{2\pi i/n}$ are allowable in an expression by radicals. Moreover, if $n = rs$, then $\sqrt[n]{b} = \sqrt[r]{\sqrt[s]{b}}$, so that at the cost of adding more steps to the chain of fields, we may assume that all the roots are pth roots, for various prime integers p. But, we note that there is a great deal of ambiguity in an expression by radicals, because there are n choices for each $\sqrt[n]{b}$.*

EXAMPLE 6.20. $(-3 + \sqrt[5]{2})^{1/4}$ *may stand for any one of 20 complex numbers, so the tower of fields*

$$\mathbb{Q} \subset \mathbb{Q}(\sqrt[5]{2}) \subset \mathbb{Q}((-3 + \sqrt[5]{2})^{1/4}$$

is not uniquely determined.

This kind of ambiguity is inherent in the notation.

PROPOSITION 6.9. *Let α be a complex number which can be expressed by radicals over F. Then its increasing chain of fields*

$$F = F_0 \subset F_1 \subset \cdots \subset F_r = K$$

satisfying the conditions (1) *and* (2) *in Definition 6.9 additionally satisfies the following property:*

(3) *for each j, F_j is a Galois extension of F_{j-1} and the Galois group $G(F_j/F_{j-1})$ is a cyclic group.*

PROOF. We consider the chain in Definition 6.9, in which

$$F_r = F(\beta_1, ..., \beta_r).$$

As we have already remarked above, we may assume that for each $j = 1, 2, ...r$, there exists a prime integer p_j such that each $\beta_j^{p_j} \in F_{j-1}$. Let $\omega_{p_j} = e^{2\pi i/p_j}$ be the p_jth root of unity, and we form a new chain of fields by adjoining the elements $(\omega_{p_1}, ..., \omega_{p_r}; \beta_1, ..., \beta_r)$ in the given order. Then by Theorem 6.4, Proposition 6.5, and Proposition 6.7, we see that each of these extensions is Galois, with cyclic Galois group. Some of the extensions in this new chain may be trivial because of redundancy. If that happens, we simply shorten the chain. Since the last field $F(\{\omega_{p_j}\}, \{\beta_j\})$ in this new chain contains F_r, it contains the given α. □

7.1. Equations with degree ≤ 4. Before we even start this section, we have seen a few examples of radical extensions of degree ≤ 4, namely, quadratic extensions, cubic extensions and quartic extensions. We have derived the formula for such equations, so we do not feel that we really have to prove the following result: As before, F is a subfield of the field $\mathbb{C}$ of complex numbers.

PROPOSITION 6.10. *Let α be a root of a polynomial $f(x)$ over F of degree ≤ 4. Then α is expressible by radicals over F.*

PROOF. For quadratic polynomials, we have the quadratic formula. For cubic equations, Cardano's formula in Subsection 3.1 gives the solution. For quartic equations, the formula in Subsection 4.2 gives the solution. □

7.2. Quintic Equations. As before, F is a subfield of the field $\mathbb{C}$ of complex numbers.

THEOREM 6.5. *Let F be a subfield of the field $\mathbb{C}$ of complex numbers. The roots of a quintic irreducible polynomial $f(x)$ over F whose Galois group is either S_5 or A_5 can not be expressed by radicals over F.*

PROOF. Let K be a splitting field of $f(x)$ over F. If $G = S_5$ then the discriminant D of $f(x)$ is not a square in F by the first part of Proposition 6.2. If this happens, then we replace F by $F(\delta)$, where δ is a square root of the discriminant D in K. Then by the second part of Proposition 6.2, the Galois group $G(K/F(\delta))$ is A_5, because its order is $|S_5|/2$. Thus, it is enough to show that the roots of $f(x) = 0$ can not be expressed by radicals over the larger field $F(\delta)$, and this reduces the case that the Galois group G is S_5 to the case that G is A_5.

So we assume that the Galois group of $f(x)$ is A_5. We recall that A_5 is a *simple* group which operates *transitively* on the set $\{1, 2, 3, 4, 5\}$. (In fact, A_n is simple if $n \geq 5$.) We will finish our proof by contradiction. For this, we suppose that there exists a root α of $f(x) = 0$ expressible by radicals over F. Say that $\alpha \in F_r$, where F_r is the end of an increasing chain of field extensions

$$F = F_0 \subset F_1 \subset \cdots \subset F_r,$$

where each extension in the chain is Galois with a cyclic Galois group, by Proposition 6.9. Since the Galois group A_5 of $f(x)$ over F is a simple group, by Proposition 5.6 we see inductively that for each i, the Galois group of $f(x)$ over F_i is also A_5 too. But, on the other hand, the polynomial $f(x)$ will not remain irreducible over F_r, since it has a root α in F_r. This means that the Galois group of $f(x)$ over F_r will not operate transitively on the set of the five roots of $f(x) = 0$ in a splitting field of $f(x)$ over F_r. In particular, the Galois group can not be the alternating group A_5, a contradiction. □

PROBLEM 3. *Is the converse this theorem true possibly with some additional conditions?*

Now, we are going to have a specific quintic irreducible polynomial $f(x)$ over $\mathbb{Q}$ whose Galois group G is S_5, so that its roots can not be expressed by radicals. The fact that 5 is prime and that the Galois group G acts transitively on the set of roots $\{\alpha_1, ..., \alpha_5\}$ impose a great limit on the possible Galois group of $f(x)$ over $\mathbb{Q}$. This is because, since the action of G on the set of 5 roots is transitive, $|G|$ is divisible by 5, and hence G should contain an element of order 5, but the only elements of order 5 in S_5 are cyclic permutations such as $\sigma = (12345)$. Under this setting, we have

LEMMA 6.3. *S_n is generated by $\sigma = (12...n)$ and $\tau = (12)$.*

PROOF. By definition, a permutation which interchanges two indices is called a *transposition.* We recall that every element of S_n is a product of transpositions, so if we can show that this is indeed the case, then we are done:

Let G be a subgroup of S_n generated by these two permutations. Then G contains $\sigma\tau\sigma^{-1} = (23)$, and hence $\sigma(23)\sigma^{-1} = (34), ...,$ and hence it contains all transpositions $(m, m+1)$. Thus then G contains $(12)(23)(12) = (13), (13(34)(13) = (14), ...,$ and hence it contains all transpositions $(1m)$. But then G contains all $(1m)(1r)(1m) = (mr)$. Therefore, $G = S_n$. □

COROLLARY 6.8. *If the Galois group G of a quintic irreducible polynomial $f(x)$ over $\mathbb{Q}$ contains a transposition, then $G = S_5$.*

PROOF. For this, we may assume that G contains the cyclic permutation $\sigma = (12345)$.Renumbering if necessary, we may assume that a transposition τ of G acts as (12). Then by Lemma 6.3, we finish the proof. □

COROLLARY 6.9. *Moreover, suppose that the irreducible polynomial in (6.10) has roots* $\{\alpha_1, ..., \alpha_5\}$, *and let* K *be its splitting field. If* $F(\alpha_1, \alpha_2, \alpha_3)$ *is a proper subfield of* K, *then* $G(K/F)$ *is the symmetric group* S_5.

PROOF. Write $F' = F(\alpha_1, \alpha_2, \alpha_3)$. We note that the only non-trivial permutation on the set $\{\alpha_1, ..., \alpha_5\}$ which fixes the three roots $\alpha_1, \alpha_2, \alpha_3$ is the transposition $\tau_4 = (45)$. Thus, if $F' \neq K$, then τ_4 must be in the Galois group $G(K/F')$ and hence $G(K/F)$ contains a transposition. Thus, by Corollary 6.8 we finish the proof. □

COROLLARY 6.10. *Let* $f(x)$ *be an irreducible quintic polynomial over* $\mathbb{Q}$ *with exactly three* real *roots,* α_1, α_2, *and* α_3. *Then its Galois group is the symmetric group* S_5, *and hence its roots can not be expressed by radicals.*

PROOF. Since $\mathbb{Q}(\alpha_1, \alpha_2, \alpha_3) \subset \mathbb{R}$ and the other two roots α_4 and α_5 are not real numbers, its splitting field K is not a subfield of $\mathbb{R}$. Thus, we conclude by Corollary 6.9 that the Galois group of $f(x)$ is S_5. Then by Theorem 6.5, the roots of $f(x)$ can not be expressed by radicals. □

LEMMA 6.4. *There are infinitely many quintic irreducible polynomials over* $\mathbb{Q}$ *whose roots are not expressed by radicals.*

PROOF. In order first to have a polynomial with exactly three real roots, we first consider

$$f(x) = x^5 - p^4 x = x(x^2 - p^2)(x^2 + p^2),$$

where p is a prime integer. This polynomial clearly has three rational integer roots, $0, \pm p$. Then the first derivative test will allow us to add a small positive constant a satisfying

$$0 < a < p/\sqrt[4]{5} \cong 0.668\,74p$$

to $f(x)$ (as a bias or an offset) in such a way that the resulting polynomial $f(x) + a$ will still have three distinct *real* roots.

Since we are also looking for the final polynomial to be irreducible polynomial over $\mathbb{Q}$, as well as having three distinct real roots, taking $a = p$ will always work for this purpose. That is, the polynomial

$$f(x) + p = x^5 - p^4 x + p$$

is irreducible over $\mathbb{Q}$ by Eisenstein Criterion (Proposition 2.11) and has three distinct *real* roots. □

EXAMPLE 6.21. *For instance, we may choose $p = 2$ to obtain an irreducible polynomial over $\mathbb{Q}$:*

$$x^5 - 16x + 2.$$

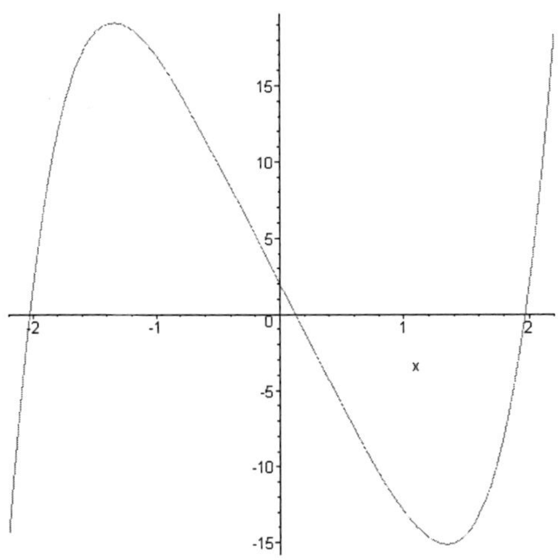

Then, by Corollary 6.10, we know that its roots can not be expressed by radicals over $\mathbb{Q}$.

EXAMPLE 6.22. *The integral polynomial*

$$f(x) = 2x^5 - 10x + 5$$

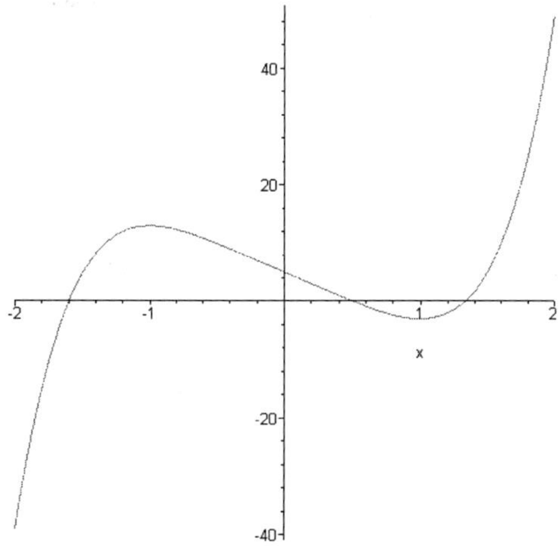

is irreducible over $\mathbb{Q}$ *by Eisenstein Criterion, and we can easily check that it has three distinct* real *roots by the first derivative test. Thus, its roots can not be expressed by radicals over* $\mathbb{Q}$.

7.3. Computation of Discriminants. In the case of irreducible polynomial equations $f(x) = 0$ over $\mathbb{Q}$ of degree ≤ 4, we have so far seen that the discriminant D of $f(x)$ was very useful in determining the Galois group of $f(x)$. Thus in what follows, we first derive the formula for discriminant D of an irreducible quintic polynomial equation $f(x) = 0$ over $\mathbb{Q}$, and try to use it in determining its Galois group. But, we already needed the resolvent cubic for $f(x)$ when $f(x)$ is an irreducible quartic polynomial over $\mathbb{Q}$, so we may expect that the discriminant alone will not determine the Galois group when $n \geq 4$, and this may require us to figure out all the transitive subgroups of S_n in advance.

Using Computer Algebra System, for instance, REDUCE, we can compute the groebner basis of the ideal generated by

$$\sigma_1 - y_1, \sigma_2 - y_2, \sigma_3 - y_3, \sigma_4 - y_4, \sigma_5 - y_5$$

according to Theorem 6.1:

---

result $:=$ *{x1 + x2 + x3 + x4 + x5 - y1,*

```
    x2**2 + x2*x3 + x2*x4 + x2*x5 - x2*y1 + x3**2 + x3*x4 + x3*x5 -
x3*y1 + x4**2 + x4*x5 - x4*y1 +
    x5**2 - x5*y1 + y2,
    x3**3 + x3**2*x4 + x3**2*x5 - x3**2*y1 + x3*x4**2 + x3*x4*x5 -
x3*x4*y1 + x3*x5**2 - x3*x5*y1 +
    x3*y2 + x4**3 + x4**2*x5 - x4**2*y1 + x4*x5**2 - x4*x5*y1 +
x4*y2 + x5**3 - x5**2*y1 + x5*y2 -
    y3,
    x4**4 + x4**3*x5 - x4**3*y1 + x4**2*x5**2 - x4**2*x5*y1
    + x4**2*y2 + x4*x5**3 - x4*x5**2*y1 + x4*x5*y2 - x4*y3
    + x5**4 - x5**3*y1 + x5**2*y2 - x5*y3 + y4,
    x5**5 - x5**4*y1 + x5**3*y2 - x5**2*y3 + x5*y4 - y5}$
```

Here,

$\sigma_1 =$ *x1+x2+x3+x4+x5,*

$\sigma_2 =$ *x1*x2+x1*x3+x1*x4+x1*x5+x2*x3+x2*x4+x2*x5*
*+x3*x4+x3*x5+x4*x5,*

$\sigma_3 =$ *x1*x2*x3+x1*x2*x4+x1*x2*x5+x1*x3*x4+x1*x3*x5*
*+x1*x4*x5+x2*x3*x4+x2*x3*x5+x3*x4*x5,*

$\sigma_4 =$ *x1*x2*x3*x4+x1*x2*x4*x5+x1*x3*x4*x5*
*+x2*x3*x4*x5,*

$\sigma_5 =$ *x1*x2*x3*x4*x5,*

are the five elementary symmetric functions in the five roots x1, x2, x3, x4,x5 of

$$f(x) = 0.$$

Then we compute:

```
result3:=preduce((x1-x2)^2*(x1-x3)^2*(x1-x4)^2*(x1-x5)^2
*(x2-x3)^2*(x2-x4)^2*(x2-x5)^2*(x3-x4)^2*(x3-x5)^2
*(x4-x5)^2,result,{x1,x2,x3,x4,x5,y1,y2,y3,y4,y5});
```

```
result3 := 256*y1**5*y5**3 - 192*y1**4*y2*y4*y5**2
- 128*y1**4*y3**2*y5**2 + 144*y1**4*y3*y4**2*y5
- 27*y1**4*y4**4 + 144*y1**3*y2**2*y3*y5**2
- 6*y1**3*y2**2*y4**2*y5 - 80*y1**3*y2*y3**2*y4*y5
+ 18*y1**3*y2*y3*y4**3 - 1600*y1**3*y2*y5**3
+ 16*y1**3*y3**4*y5 - 4*y1**3*y3**3*y4**2
+ 160*y1**3*y3*y4*y5**2 - 36*y1**3*y4**3*y5
- 27*y1**2*y2**4*y5**2 + 18*y1**2*y2**3*y3*y4*y5
- 4*y1**2*y2**3*y4**3 - 4*y1**2*y2**2*y3**3*y5
+ y1**2*y2**2*y3**2*y4**2 + 1020*y1**2*y2**2*y4*y5**2
+ 560*y1**2*y2*y3**2*y5**2 - 746*y1**2*y2*y3*y4**2*y5
+ 144*y1**2*y2*y4**4 + 24*y1**2*y3**3*y4*y5
- 6*y1**2*y3**2*y4**3 + 2000*y1**2*y3*y5**3
- 50*y1**2*y4**2*y5**2 - 630*y1*y2**3*y3*y5**2
+ 24*y1*y2**3*y4**2*y5 + 356*y1*y2**2*y3**2*y4*y5
- 80*y1*y2**2*y3*y4**3 + 2250*y1*y2**2*y5**3
- 72*y1*y2*y3**4*y5 + 18*y1*y2*y3**3*y4**2
- 2050*y1*y2*y3*y4*y5**2 + 160*y1*y2*y4**3*y5
- 900*y1*y3**3*y5**2 + 1020*y1*y3**2*y4**2*y5
- 192*y1*y3*y4**4 - 2500*y1*y4*y5**3 + 108*y2**5*y5**2
- 72*y2**4*y3*y4*y5 + 16*y2**4*y4**3 + 16*y2**3*y3**3*y5
- 4*y2**3*y3**2*y4**2 - 900*y2**3*y4*y5**2
+ 825*y2**2*y3**2*y5**2 + 560*y2**2*y3*y4**2*y5
- 128*y2**2*y4**4 - 630*y2*y3**3*y4*y5
+ 144*y2*y3**2*y4**3 - 3750*y2*y3*y5**3
+ 2000*y2*y4**2*y5**2 + 108*y3**5*y5
- 27*y3**4*y4**2 + 2250*y3**2*y4*y5**2
- 1600*y3*y4**3*y5 + 256*y4**5 + 3125*y5**4$
```

```
result4:=sub({y1=0,y2=0,y3=0},result3);
result4:=256*y4**5 + 3125*y5**4$
```

This shows that the discriminant

$$\begin{aligned} D \;=\; & (x1-x2)^2(x1-x3)^2(x1-x4)^2(x1-x5)^2 \\ & \cdot(x2-x3)^2(x2-x4)^2(x2-x5)^2 \\ & \cdot(x3-x4)^2(x3-x5)^2 \\ & \cdot(x4-x5)^2 \end{aligned}$$

is indeed a symmetric polynomial by the first part of Theorem 6.1. Moreover, by the second part of Theorem 6.1, we see that

$$\begin{aligned} D \;=\; & +256\sigma_1^5\sigma_5^3 - 192\sigma_1^4\sigma_2\sigma_4\sigma_5^2 - 128\sigma_1^4\sigma_3^2\sigma_5^2 + 144\sigma_1^4\sigma_3\sigma_4^2\sigma_5 \\ & -27\sigma_1^4\sigma_4^4 + 144\sigma_1^3\sigma_2^2\sigma_3\sigma_5^2 - 6\sigma_1^3\sigma_2^2\sigma_4^2\sigma_5 - 80\sigma_1^3\sigma_2\sigma_3^2\sigma_4\sigma_5 \\ & +18\sigma_1^3\sigma_2\sigma_3\sigma_4^3 - 1600\sigma_1^3\sigma_2\sigma_5^3 + 16\sigma_1^3\sigma_3^4\sigma_5 - 4\sigma_1^3\sigma_3^3\sigma_4^2 \\ & +160\sigma_1^3\sigma_3\sigma_4\sigma_5^2 - 36\sigma_1^3\sigma_4^3\sigma_5 - 27\sigma_1^2\sigma_2^4\sigma_5^2 + 18\sigma_1^2\sigma_2^3\sigma_3\sigma_4\sigma_5 \\ & -4\sigma_1^2\sigma_2^3\sigma_4^3 - 4\sigma_1^2\sigma_2^2\sigma_3^3\sigma_5 + \sigma_1^2\sigma_2^2\sigma_3^2\sigma_4^2 + 1020\sigma_1^2\sigma_2^2\sigma_4\sigma_5^2 \\ & +560\sigma_1^2\sigma_2\sigma_3^2\sigma_5^2 - 746\sigma_1^2\sigma_2\sigma_3\sigma_4^2\sigma_5 + 144\sigma_1^2\sigma_2\sigma_4^4 + 24\sigma_1^2\sigma_3^3\sigma_4\sigma_5 \\ & -6\sigma_1^2\sigma_3^2\sigma_4^3 + 2000\sigma_1^2\sigma_3\sigma_5^3 - 50\sigma_1^2\sigma_4^2\sigma_5^2 - 630\sigma_1\sigma_2^3\sigma_3\sigma_5^2 \\ & +24\sigma_1\sigma_2^3\sigma_4^2\sigma_5 + 356\sigma_1\sigma_2^2\sigma_3^2\sigma_4\sigma_5 - 80\sigma_1\sigma_2^2\sigma_3\sigma_4^3 \\ & +2250\sigma_1\sigma_2^2\sigma_5^3 - 72\sigma_1\sigma_2\sigma_3^4\sigma_5 + 18\sigma_1\sigma_2\sigma_3^3\sigma_4^2 \\ & -2050\sigma_1\sigma_2\sigma_3\sigma_4\sigma_5^2 + 160\sigma_1\sigma_2\sigma_4^3\sigma_5 - 900\sigma_1\sigma_3^3\sigma_5^2 \\ & +1020\sigma_1\sigma_3^2\sigma_4^2\sigma_5 - 192\sigma_1\sigma_3\sigma_4^4 - 2500\sigma_1\sigma_4\sigma_5^3 + 108\sigma_2^5\sigma_5^2 \\ & -72\sigma_2^4\sigma_3\sigma_4\sigma_5 + 16\sigma_2^4\sigma_4^3 + 16\sigma_2^3\sigma_3^3\sigma_5 - 4\sigma_2^3\sigma_3^2\sigma_4^2 \\ & -900\sigma_2^3\sigma_4\sigma_5^2 + 825\sigma_2^2\sigma_3^2\sigma_5^2 + 560\sigma_2^2\sigma_3\sigma_4^2\sigma_5 - 128\sigma_2^2\sigma_4^4 \\ & -630\sigma_2\sigma_3^3\sigma_4\sigma_5 + 144\sigma_2\sigma_3^2\sigma_4^3 - 3750\sigma_2\sigma_3\sigma_5^3 \\ & +2000\sigma_2\sigma_4^2\sigma_5^2 + 108\sigma_3^5\sigma_5 - 27\sigma_3^4\sigma_4^2 + 2250\sigma_3^2\sigma_4\sigma_5^2 \\ & -1600\sigma_3\sigma_4^3\sigma_5 + 256\sigma_4^5 + 3125\sigma_5^4 \,. \end{aligned}$$

is the discriminant D of an *irreducible* polynomial $f(x)$ over $\mathbb{Q}$:

$$f(x) = x^5 - \sigma_1 x^4 + \sigma_2 x^3 - \sigma_3 x^2 + \sigma_4 x - \sigma_5.$$

EXAMPLE 6.23. *Thus, if* $f(x) = x^5 - 16x + 2$, *then* $\sigma_1 = 0, \sigma_2 = 0, \sigma_3 = 0$, *and hence the discriminant of* $f(x)$ *is*

$$\begin{aligned} D &= 256\sigma_4^5 + 3125\sigma_5^4 \\ &= 256 \cdot (-16)^5 + 3125 \cdot (-2)^4 \\ &= -268385456, \end{aligned}$$

is not a square in $\mathbb{Q}$, *as we expected. In fact, by part a of Proposition 6.2 the Galois group* $G = G(K/\mathbb{Q})$ *is not a subgroup of the alternating group* A_5. *But, it could be a proper subgroup of* S_5.

On the other hand, since it has exactly two complex roots, it contains a transposition and hence it must be S_5. *Thus, its roots can not be expressed by radicals.*

EXAMPLE 6.24. *If* $f(x) = x^5 - 5x + \frac{5}{2}$ *over* $\mathbb{Q}$, *then its discriminant*

$$\begin{aligned} D &= 256\sigma_4^5 + 3125\sigma_5^4 \\ &= 256 \cdot (-5)^5 + 3125 \cdot (\frac{-5}{2})^4 \\ &= (-10846875)/16, \end{aligned}$$

is not a square in $\mathbb{Q}$, *as we expected.*

Thus, from these two examples we see that if the equation has three distinct real roots, we do not even need to compute the discriminant D of $f(x)$. Otherwise, it will still provide a clue for finding the Galois group of $f(x)$ in almost all cases:

EXAMPLE 6.25. *If* $f(x) = x^5 - 5x + 10$ *is the irreducible polynomial over* $\mathbb{Q}$, *then its discriminant*

$$\begin{aligned} D &= 256\sigma_4^5 + 3125\sigma_5^4 \\ &= 256 \cdot (-5)^5 + 3125(-10)^4 \\ &= 97656250256/3125 \\ &= (2^4 \cdot 3049 \cdot 2001809)/5^5 \end{aligned}$$

is not a square in $\mathbb{Q}$. *Thus by part a of Proposition 6.2 the Galois group* $G = G(K/\mathbb{Q})$ *is not a subgroup of the alternating group* A_5. *But, it could be still a proper subgroup of* S_5, *not containing a transposition but containing an odd permutation of order* > 2, *since it is not a subgroup of* A_5.

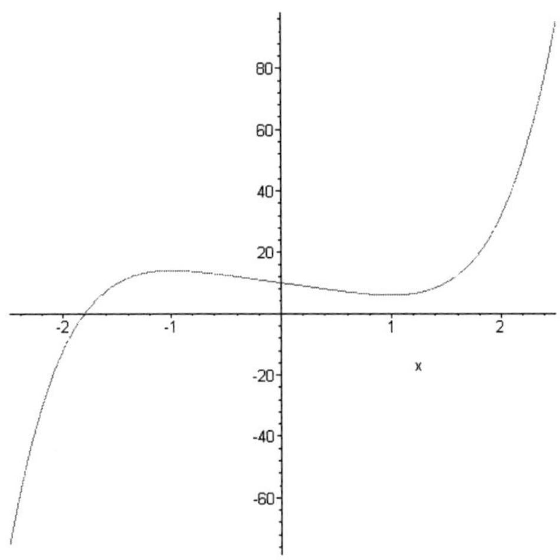

We note that this polynomial has only one real root. In fact, the following lines for REDUCE say that its 5 numerical solutions and only one of them is a real number:

———————

```
on fullroots;
on rounded;
solve(x**5 - 5*x + 10);
{x=1.22140106699 + 0.650175291569*i,
x=1.22140106699 - 0.650175291569*i,
```

*x= - 0.320305635197 + 1.67201504509*i,*
*x= - 0.320305635197 - 1.67201504509*i,*
x= - 1.80219086358}$

Exercises for Section 6.7

EXERCISE 6.24. *Determine the transitive subgroups of* S_5.

EXERCISE 6.25. *Let* G *be the Galois group of an irreducible quintic polynomial. Prove that if* G *contains an element of order* 3*, then* G *is either* S_5 *or* A_5.

EXERCISE 6.26. *Can the roots of the* irreducible *polynomial* $f(x) = x^5 - 5x + 10$ *over* $\mathbb{Q}$ *in Example 6.25 be expressed by radicals over* $\mathbb{Q}$*? Is its Galois group Klein four group or not? (Hint: We recall that every finite p-group is solvable, and Klein four group is an abelian group, so it is solvable.)*

EXERCISE 6.27. *Since any finite p-group is solvable, we only consider subgroups of* S_5 *which are not a p-group. Among such subgroups of* S_5 *of orders,*

$$\begin{aligned}&3 \cdot 5,\\&2^3 \cdot 3, 2^2 \cdot 3, 2 \cdot 3,\\&2^3 \cdot 5, 2^2 \cdot 5, 2 \cdot 5,\\&2^3 \cdot 3 \cdot 5,\end{aligned}$$

which are not solvable? We already know that S_5 *is not solvable. Think about those subgroups of order* $< 2^3 \cdot 3 \cdot 5$ *listed above which moreover contain Klein four group* V *as its subgroup.*

EXERCISE 6.28. *Is every Galois extension of degree* 10 *expressible by radicals?*

APPENDIX A

An Introduction to Algebraic Analysis

1. Tools in Mathematics

The whole body of Mathematics consists of many different kinds of knowledge, or results and procedures which produce such results, most of which are verified through the process of the so-called "proof". Those proved and somewhat useful or important are usually called **theorems**. Unproven ones are formally called **conjectures**. The place for most mathematicians should be between these two words, because we are born to establish reasonable conjectures and then prove them. When one tries to prove something, something itself must be first recognized as a reasonable mathematical object; discovering such object could be quite personal.

In proving reasonable conjectures, one generally uses two tools, a special type of logics, called **mathematical logic**, and **computation,** definitely guided by his own mathematical intuitions or experiences. Since they are applied as tools in the process of the so-called **proof**, they themselves cannot be called mathematics. However, the kinds of the methods employed certainly put some unavoidable limitations on the character of the consequences. These two tools, nevertheless, do not hurt each other. In contrast, they help each other, and these two are the most powerful tools we, mathematicians, have. Depending on the degree of putting more emphasis on any of these two, one usually has two basic tendencies in doing mathematics, axiomatic or computational.

1.1. Axiomatic Nature. Although it is very difficult to figure out what constitutes mathematics, one may describe its axiomatic nature as follows: Once a reasonable mathematical conjecture is recognized, one accepts, without questioning their existence or validity, a finite number of undefined terminologies and that of postulates (i.e., axioms or principles) which describe some relations between the undefined terminologies, and one applies a finite number of steps of mathematical logic directly to them or to some of their consequences until one finally reaches the desired positive or negative conclusions. As a demonstrative science started even before Greek period, mathematics has shown this axiomatic nature for a long time. One can find such an evidence in, for instance, Euclid's Elements [**E**] and other works of Greek mathematicians and philosophers. In fact, one can even trace back to their works to see that most of the formal terminologies used in mathematics, something like definitions, propositions, corollaries, etc., were already used in his Elements and other Greek literatures. The meaning and usage of these terminologies and other metamathematical concerns are also well explained in the introduction of A. Whitehead and B. Russell's famous work, *Principia Mathematica* [**WR**], which was originally published in 1910.

1.2. Computational Nature. When computations in general are concerned, we know that there are basically two different kinds, namely, exact computation and approximation. **Approximation** is done through the limiting process mostly in terms of inequalities, where as **exact computation or symbolic computation** is done through the so-called four binary operations: addition, subtraction, multiplication and acceptable division. These operations are applied not only to numbers but also to a mixture of numbers and **symbols**.

Computations are very important to us at least for the following three reasons. First, they usually show up in the middle of the process called proof. This is almost unavoidable to every mathematician. Secondly, a decent amount of computational experiments quite often reveals to us what's going on in a very clean manner, while simple-minded 'what if ...' ways of logical questioning cannot. Thirdly, we, mathematicians, can communicate with other branches of sciences very efficiently and successfully, through computational results and their usefulness. Considering that mathematics has so far grown up to this much within the accumulation of our civilization, this point of view is also very important, because isolation only hurts oneself till one's starvation.

Up to the end of the 19th century or the beginning of the 20th century, some important mathematical works discovered and proved were based on individual mathematician's heavy hand-computational experiments. However, as was the case in Invariant Theory of the 19th century, mathematicians were practically forced out to give up heavy computations when the number of unknowns and the size of numbers are substantially increased so that hand-calculation obviously becomes terrible. Any attempt based on computational experiments may be called a **computational approach**.

2. Overemphasis on Axiomatic Approach in the 20th Century

The revival of the so-called **axiomatic approach** after the Greek period (namely, Euclid's Elements) is clearly evident in the Lecture Notes of E. Artin and E. Noether, written up by B.L. van der Waerden in 1930s [**BW**]. This approach, especially well-supported by precise notations adapted from the subject, Set Theory, was so good in pursuing a particular-minded concept or property, that it rapidly pushed out the old tradition of computational approaches, which was quite common up to the first quarter of the 20th century, from the mathematical world. In addition, some eminent mathematicians, like Banach in Analysis or later E. Spanier in Algebraic Topology [**ES**], tried to figure out and establish sufficiently small number of axioms but still good enough to explain successfully the rest of the theory they know. In order to fulfill such tasks, they examined many computed examples of different kinds and pulled out some common underlying axioms which replace too much involved terrible computations. The success of this attempt is quite evident in almost every part of current mathematics in both research articles and textbooks. It is also around this period that most mathematicians officially but still reluctantly began to acknowledge the usefulness of the non-traditional mathematical axiom, i.e., the axiom of choice, or its equivalence, the Zorn's Lemma.

The point of views expressed in this article could be quite personal until they get more acceptance from general audience. However, the readers will not be hurt by the author's viewpoints on mathematics expressed in this article but it will supply a fresh taste or sense to our main meal, mathematics.

2.1. Danger of Too Much Emphasis on Axiomatic Approach. One should note that those mathematicians, who introduced an appropriate set of axioms for the knowledge they already know, were also very good at computation, since they were trained as, called from now on, the "first" generation mathematicians, by their teaches most of whom were under the strong influence of computational approach of the 19th century. On the other hand, the "second" generation mathematicians, i.e., those who were mathematically trained by the first generation mathematicians, tended to start their mathematics not from heavy computational experiments but from a small amount of axioms, and they were more and more away from computational experiences. The lack of computational experiences is more significant in the "third" generation mathematicians, i.e., those trained by the second generation mathematicians.

Nevertheless, the strength of axiomatic approach had been fully demonstrated by the monumental works of a single mathematician, Alexander Grothendieck (1928 -), on the Foundation of Algebraic Geometry, [**AG1**], FGA [**AG2**], EGA's [**AG3**] and [**AG4**], and SGA's [**AG5**] during 1950s up until the beginning of 1970s. His works in turn simply recall the works of a group of mathematicians, called Bourbaki [**B2**] and [**B1**], as elementary results. Anyone who is very much interested in the scope of axiomatic approach in mathematics explored during the 20th century is strongly encouraged to study and digest the above mentioned Grothendieck's works, of which the basic concepts and results are described in terms of categorical languages, called the **language of schemes**.

In particular, the tendency of the results obtained under a strong influence of axiomatic approach is too much conceptual, and hence becomes more and more **soft** and hence fragile, as time goes. In this case, a good-minded mathematician should worry about whether his and/or her results will still survive after, say, $1,000$ years from now. A prototypical example is Euclid's algorithm for computing the greatest common divisor of two numbers. His method is still known to us because it is about computation of certain mathematical objects in an efficient manner. Moreover, lack of computability only speeds up the isolation of Mathematics from other branches of Sciences. This is in no way acceptable to us.

For an elegant axiomatic treatment but still with a strong emphasis on computational point of views, of Algebraic Geometry, one may refer to Prof. George R. Kempf's two books [**GK1**] and [**GK2**].

3. A New Computational Tool: Computers

3.1. Personal Computers and High Performance Servers. Starting at around the beginning of the fourth quarter of the 20th century, computer industry has grown up with an unexpected rate of success; this nowadays allows us to compute very big numbers using a Personal Computer or even a Notebook Computer. Recently, inexpensive High Performance Cluster Servers are also available to us. As a tool for actual computation, computers are already playing a significant role and such role will surely change the mathematical world of the 21-st century. Here are some examples.

EXAMPLE A.1. *Here, $p(n)$ denotes, by definition, the number of ways of writing a given natural number n as a sum of others (called components) ignoring the order among the components. One starts with a very famous theorem:*

THEOREM A.1. *(Rademacher* [**GA**]*) Given a natural number n, one has*

$$p(n) = \frac{1}{\pi\sqrt{2}} \sum_{k=1}^{\infty} A_k(n)\sqrt{k}\frac{d}{dn}\frac{\sinh((\pi/k)\sqrt{(2/3)(n-\frac{1}{24})})}{\sqrt{n-\frac{1}{24}}}, \tag{A.1}$$

where

$$A_k(n) = \sum_{\substack{0 \le h < k \\ (h,k)=1}} \varpi_{h,k} \exp(-2\pi inh/k),$$

where

$$\varpi_{h,k} = \exp(\pi is(h,k)),$$

is a $24k$th root of unity and where

$$s(h,k) = \sum_{\mu=1}^{k-1} \left(\frac{\mu}{k} - \left[\frac{\mu}{k}\right] - \frac{1}{2}\right)\left(\frac{h\mu}{k} - \left[\frac{h\mu}{k}\right] - \frac{1}{2}\right)$$

is the so-called Dedekind sum.

It turns out that the value of $p(20,000,000)$, when computed it using this theorem, requires more than $4,900$ number of decimal digits (Cf. [**STS1**]*). Thus the final result is really a super-ultra big number, which is one point. Moreover, another point is that it only took about 17 Minutes with already-done pre-calculations of the so-called Dedekind sums, under the same hardware and software limitations as in the next example. These two points clearly show us the computing capability of current PC's.*

REMARK A.1. *One should note that computation of such a super-ultra number eventually raises a serious concern whether the computed result is absolutely correct. This is certainly a new kind of headaches to mathematicians.*

On the other hand, the author recently discovered an exact algebraic formula for a graded partition function $p_{\mathbf{M}}(n)$ over a multiset $\mathbf{M}$ and an associated quasi-recursive formula for $p_{\mathbf{M}}(n)$, both of which never *require approximation at all (Cf.* [**STS2**]*). One should note that it is neither quite* *possible to obtain an analytically exact formula for* $p_{\mathbf{M}}(n)$ *nor to obtain a recursive formula for it, something like either Rademacher's formula or the well-known Euler's recursive formula for* $p(n)$*, unless there is a significant symmetry among the elements of* $\mathbf{M}$. *However, one version of the author's formulas computes all the values of* $p_{\mathbf{M}}(i)$ *for* $1 \le i \le n$ *with time efficiency of* $\mathrm{O}(n^2)$ *for a general multiset* $\mathbf{M}$ *(Cf.* [**STS3**]*).*

3.2. Computer Algebra Systems. Of course, having just a very fast computing machine of which hardware is very reliable is not enough. One also needs a reliable software which guarantees several facilities. Among them, the most urgent requirements are, for instance, ability to handle numbers of any sizes and arbitrary precision in representing fractions, each of which absolute reliability is definitely required in any computation with computers.

EXAMPLE A.2. *The following Table shows the actual computation of the value of the classical partition function* $p(n)$*, for* $n = 200$*, using the Rademacher's formula in Eq (A.1). Here, by adding the first* 8 ***non-zero*** *terms of fractional real numbers, one gets a fractional real number,*

$$3972999029388.001346332,$$

which is very close to the true integer value $p(200) = 3972999029388$.

$$\begin{array}{rcr}
t(1) & = & +3.97299899318589575787 2E+12 \\
t(2) & = & +36282.9784854411847056 \\
t(3) & = & -87.58391287399 \\
t(4) & = & +5.146534584 \\
t(5) & = & +1.424217569 \\
t(6) & = & +0.07102143214 \\
t(7) & = & 0 \\
t(8) & = & +0.04357655366 \\
t(9) & = & +0.0256657539 \\
{[t(10)} & = & +0.008603641042] \\
\hline
\text{Up to 9th Partial Sum} & = & +3972999029388.001346332
\end{array}$$

When it is about computation of an exact integer value in this way of approximation, there should not be any approximation-related errors in all of the intermediate steps. Do not expect that an error-free computation of such a kind is quite easy (Cf. [**STS1**]*), unless one uses very reliable Computer Algebra Systems.*

REMARK A.2. *This example, which is prepared by the author of this book, is especially interesting because it is a direct computational verification of the famous example computed by Hardy and Ramanujan (p. 284 of* [**HR**]*) and quoted by Andrews (p. 70 of* [**GA**]*). Indeed, the original calculation in Hardy's own handwriting occurs on the dust jacket of the 1969 edition of "A Mathematician's Apology",* [**H**]. *To our surprise, the calculation by Hardy and Ramanujan contains an error. The corrected calculation as in Example A.2 turns out to be much closer to* $p(200)$ *than the original incorrect version. Therefore, the third term,* -87.555*, in the original computation table for* $p(200)$ *in p. 284 of* [**HR**] *and later quoted in p. 70 of* [**GA**]*, needs to be replaced with* -87.584.

Any software package lacking these two capabilities should not be our choice for a tool for computational experiments. In addition to them, it is also required to have an ability to solve a given system of polynomial equations, i.e., non-linear equations over, say, the field of complex numbers. This means that it must have an ability to handle symbols as well as numbers. The theory for such computation had not fully understood until Bruno Buchberger (1942 -) discovered his famous algorithm for Groebner bases in 1965. Any reliable software package with at least these three facilities could be called a **Computer Algebra System**.

There are two types of Computer Algebra Systems, commercially available or not. Some of the commercially available ones are, MAPLE, MACSYMA, REDUCE, etc. Some of non-commercially available ones are, MACAULAY 2, CoCoA, etc. There should be more than these.

3.3. Behind Computational Experiments. One first notes that there are two kinds of computations one can do with computers, **exact computation** and **approximation**. Exact computation only needs addition, subtraction, multiplication, and allowed division depending on given conditions, on a given mixture of known symbols, unknown symbols, and numbers. In contrast to this, approximation allows us to apply non-exact division and limiting process, as well as addition, subtraction, multiplication and division on given numbers only. Thus, exact computation allows us to manipulate not only numbers but also **symbols**, whereas approximation only allows us to manipulate numbers. This tells us why exact computation is also called **Algebraic Analysis** or **Symbolic Analysis**, while approximation is called **Numerical Analysis**. Examples of Numeric Analysis are numerous. For instance, FEM (Finite Element Method) is just one of them. Also, see Example A.2 above.

These two approaches in computational methods do not hurt each other. In fact, they help each other. The only general trouble is that Algebraic Analysis does require some amount of elementary Algebra, mostly about polynomial rings in more than one unknowns, as well as some amount of Modern Algebraic Geometry, developed by Oscar Zariski [**OZ**] and J. P. Serre [**JS**], and a few others (For beginners, this amount of accounts on the subject may be good enough).

A proper sequence of computational approach is as follows:

(1) to apply appropriate exact methods in Algebraic Analysis until all available ones are exhausted, and then, as a final step,
(2) to apply appropriate approximation methods in Numeric Analysis.

The subject of Numeric Analysis has been well known to us after Sir Isaac Newton. However, Algebraic Analysis is very new to us. The most important theoretical background for recent revival of computational approach is the Buchberger's astonishing discovery of the algorithm for Groebner bases for polynomial ideals in more than one indeterminates. He figured out the method while he was working on his Ph.D. thesis, and it was first recognized by mathematicians oriented to computer science somewhat before or around 1975. I myself learned this remarkable algorithm in mid 1980s.

3.4. Danger of Too Much Emphasis on Computational Approach. If one puts too much emphasis on computational approach, he will soon find that he does not know why the method works and eventually he will get lost, because the arithmetic operations themselves never tell him what's going on, unless he knows *in advance* what's really going on. Unfortunately, it is obviously impossible most of the time, especially when one works on some mathematical phenomena not yet well understood to him. This means that one really needs a well-developed theory for exact computation or Algebraic Analysis.

There was no general theory for exact computation with computers until B. Buchberger discovered a very important algorithm. The mixture of his algorithm with the division algorithm produces very important algorithms for polynomial ideals in more than one variables. It is very urgent to understand this theory for hard but honest and diligent computation. The readers of this article should not underestimate the strength and usefulness of this theory.

4. Expected Trends of the 21st Century Mathematics

4.1. Computer Age. It is well known that there were two types of computers, analogue and digital, in the past. Nowadays, however, computers usually mean **digital computers** or their variants for better performance, since performance of digital computers (for instance, signal to noise ratio) has already proved to be a lot better than analogue types. It is also well known that the theoretical creation of digital computers was done by a famous mathematical logician, von Neumann, who invented Class Theory to answer, for instance, the question like whether the set of all sets is a set or not. After theoretically creating digital computers, he immediately developed a good deal of approximation theory for computing the values of non-polynomial functions with his theoretical machines.

Although Electric Engineering, Computer Engineering, and Computer Science are respectively needed to establish hardware, architecture, and operating system and software, the final combination of them is a system, called, computer, and the main reason for developing such a system is to compute or calculate something in any possible senses. The point we, mathematicians, should not miss is that **computation belongs to mathematics**. Thanks to computers, paperless computation of huge numbers is already available to those who know how to use some of the commercially or non-commercially available Computer Algebra Systems, so that the scope of Algebra and Algebraic Geometry in both pure and applied directions has already been expanded a lot.

4.2. Computers Are for Computation Related Human Activities. Since computation solely belongs to mathematics, it is mathematician's due obligation to examine and develop the needed mathematical algorithms in order to obtain from the machine the desired results with absolute correctness. In addition to this, our second obligation is to make it as efficient as possible. If it takes too much computation time, then such a computational approach will never buy other people's approval at all. To my regret, these two can't be easily accomplished by those who were not trained as mathematicians.

Unlike any other centuries, existence of so many mathematical objects have been proved during the 20th century, mostly thanks to active application of axiomatic methods influenced by Set Theory or Category Theory. These results are all inputs for further investigation that which of them are computationally constructible, applying the methods of Algebraic Analysis.

I should mention one dangerous aspect of the Zorn's Lemma or its equivalent axioms. If one used the Zorn's Lemma in some place of his proof of the existence of certain mathematical object, then his proof is computationally useless in the sense that it never really tells us how to construct such an object only using the four arithmetic operations and the limiting process. This is really a disappointing aspect of the Zorn's Lemma, but is, as long as **algorithmic construction** is concerned, still important enough to keep in mind in order for us not to use it in our proof as possible as we could.

Computers have already become a convenient and very capable tool for computations in any part of mathematics. We do not necessarily require to have an access to Supercomputers, Servers, or even Workstations for computational experiments. We only need to have Personal Computers for most computational experiments. Since Personal Computers are nowadays available to everyone and since they are already too fast and too capable of, most computational experiments could be done with them without spending too much time and money. Moreover, the hardware-related performance of these machines are nowadays very much reliable. The author strongly believes that such a very reliable computational environment, already available to everybody, sooner or later changes the whole atmosphere of mathematics of the 21-st century. For instance, the so-called **internet computing** will soon become very common.

One of the main ingredients of mathematics is perhaps its **exactness** in the process of both **proof** and **computation.** The point is that both of them require quite different types of exactness. This may be also the main reason why no other branches of Sciences even try to compete with mathematics, but only to rely on the results produced by mathematicians. Consequently mathematics has its own everlasting life. We should not give up any of these two powerful tools, proof and computation, to other branches of Sciences. On the other hand, we have been silently forced to ignore the significance of computational aspects for at least a century, mostly because of our inability in exactly hand-computing mixtures of big numbers and many parameters and also partly because of strong influences of axiomatic approach of the 20th century. Isn't it a time for us to revive computation related mathematical activities with a new tool, computers, to balance ourselves between them?

5. About Our InetCompu Service

We have recently started an E-mail based Internet Computing Service, called InetCompu.

If one visits

http://trinitas.mju.ac.kr/intro2InetCompu.html

the detailed information and several demonstrative services are provided:[1]

5.1. Our Concept and Terminology. If a person, for instance, using PC, PDA, or Mobile Phone, can perform the kinds of computations he wants on a remote Server, where he is not even required to have his own account, and is also allowed to get back the computed results immediately, then we call such a service, **InetCompu** (Internet Computing, Copyrighted) or **GridCompu** (Grid Computing, Copyrighted).

5.2. What We Have Done So Far. Several years ago, the author of this book figured out a practical method to actually realize the needed Software Interface (called General Purpose Grid Middleware Interface) for such an InetCompu Service, and he finished the development of the first primitive but working version of the Interface at around August 16, 2001. Later on, the Interface for InetCompu was significantly upgraded by him to such a degree that it could handle the in-comming data more neatly and smartly. The final version currently in usage also passed a rigorous local fields test successfully.

As of August 28, 2001, we began to provide an InetCompu Service, i.e., whoever sends an E-mail to a particular E-mail address of our Server, for example, containing

(1) either an input data for an application program in our Server for a certain mathematical formula (for instance, input: n=1234567 to numbpart@trinitas.mju.ac.kr in plain text format

(2) or his own program, for instance, written in Reduce commands,

then he will soon receive the computed results from our Server.

[1]http://ch.catholic.or.kr/pundang/archive/4/intro2InetCompu.html

5.3. An Example of InetCompu. In particular,

http://trinitas.mju.ac.kr/intro2numbpart.html

explains how one can use our InetCompu Service to compute the value of $p(n)$ for a given input n, which utilizes Rademacher's formula for $p(n)$ in Theorem A.1.[2]

[2]http://ch.catholic.or.kr/pundang/archive/4/intro2numbpart.html

Bibliography

[AL] W.W. Adams and P. Loustaunau, *An Introduction to Gröbner Bases,* GSM Vol. 3, AMS, 1994.

[GA] G.E. Andrews, *The Theory of Partitions,* Reading, 2nd printing,1981, Addison-Wesley

[BW] T. Becker and V. Weispfenning, *Groebner Bases: A Computational Approach to Commutative Algebra*, Springer Verlag, 1993.

[B1] N. Bourbaki, *Algebra,* Part I and II (in English), Springer-Verlag, 1970, 1981.

[B2] N. Bourbaki, *Commutative Algebra* (in English), Springer-Verlag, 1985.

[CLO] D. Cox, J. Little, and D. O'Shea, *Ideals, Varieties, and Algorithms, An Introduction to Computational Algebraic Geometry and Commutative Algebra,* 2nd Ed., Springer-Verlag, 1997.

[ER] D. Eisenbud and L. Robbiano, *Computational Algebraic Geometry and Commutative Algebra,* Symposia Mathematica Vol. XXXIV, Istituto Nazionale di Alta Mathematica Francesco Severi, Cambridge Univ. Press, 1993.

[DE] D. Eisenbud, *Commutative Algebra with a View Toward Algebraic Geometry,* Springer-Verlag, 1995.

[E] Euclid, *The Elements,* Vol. 1 - 3 (English translation), 2nd Ed., Dover, 1956.

[AG1] A. Grothendieck, *Algébre Homologique,* Secrétariat mathématique, 11 rue Pierre Curie, PARIS 5e, 1958.

[AG2] A. Grothendieck, *Fondements de la Géométrie Algébrique* (Extracted from Bourbaki Seminar 1957-1962), Secrétariat mathématique, 11 rue Pierre Curie, PARIS 5e, 1962.

[AG3] A. Grothendieck and J.A. Dieudonné, *Eléments de Géométrie Algébrique,* Vol. I-IV, IHES, 1960-1967.

[AG4] A. Grothendieck and J.A. Dieudonné, *Eléments de Géométrie Algébrique I,* Springer-Verlag, 1970.

[AG5] A. Grothendieck, *Séminaire de Géométrie Algébrique,* Vol. 1-7, IHES, 1960-1962.

[HR] G. H. Hardy and S. Ramanujan, *Asymptotic Formulae in Combinatory Analysis*, Proc. London Math. Soc. (2), Vol. 17 (1917), pp. 75-115. (Reprinted in

Collected Papers of S. Ramanujan, pp. 276-309, Cambridge University Press, London, 1927.)

[H] G.H. Hardy, *A Mathematician's Apology*, Cambridge University. Press, 1969.

[GK1] G. Kempf, *Algebraic Varieties*, London Math. Soc. LNS 172, Cambridge Univ. Press, 1993.

[GK2] G. Kempf, *Algebraic Structures,* Friedrich Vieweg & Sohn, 1995.

[DM] D. Mumford, *Algebraic Geometry I, Complex Projective Varieties,* Springer-Verlag, 1976.

[JS] J. P. Serre, *Faisceaux Algébriques Cohérents*, Annals of Mathematics, Vol 61, No. 2, March, 1955.

[STS1] S.T. Soh, *A Reliable Implementation of the Formula of Rademacher for the Partition Function* $p(n)$, preprint

[STS2] S.T. Soh, *A Diophantine Construction of an Exact Algebraic Formula for Graded Partition Functions*, J. Korean Math. Soc., Vol 36 (1999), No. 2, pp. 267-298

[STS3] S.T. Soh, *Efficient (Quasi-) Recursive Formulas for Graded Partition Functions,* preprint

[ES] E.H. Spanier, *Algebraic Topology,* Springer-Verlag, 1966.

[BS] B. Sturmfels, *Algorithms in Invariant Theory,* Texts and Monographs in Symbolic Computation, Springer-Verlag, 1993.

[BW] B.L. van der Waerden, *Algebra,* Vol. 1 and 2, Frederick Ungar Pub., 1970.

[WR] A.N. Whitehead and B. Russell, *Principia Mathematica*, Cambridge Univ. Press, 1910.

[OZ] O. Zariski, *Collected Papers,* Vol. 1 - 4., MIT Press, 1972.

[ZP] O. Zariski and P. Samuel, *Commutative Algebra,* Vol.1 and 2, Springer-Verlag, 1958.

APPENDIX B

Computer Algebra Systems

In recent years, several Computer Algebra Systems were developed and they are currently used by many scientists and engineers. Among them, MAPLE, MATHEMATICA, REDUCE, MACSYMA, MATLAB etc. are commercially available. In this appendix, we first discuss the Groebner basis package in REDUCE, and then we discuss the Linear Algebra package of MAPLE.

1. The Groebner Basis Package in REDUCE

REDUCE is another well-known commercially available Computer Algebra System(CAS). To load a package for Groebner basis computation, one needs to type in the following line at Reduce prompt:

load_package groebner; (and then press Enter key)

REMARK B.1. *Here are quick explanations of some of those commands we have used in our codes for the REDUCE throughout this book:*

; displays the result of the command.
$ does not display the result of the command.
bye$ finishes the current reduce session.

% begins a one line comment.
comment begins a multiple line comment until ; or $

---

sqrt(2) displays square root of 2 knowing that its square is 2.
i displays $\sqrt{-1}$ knowing that its squre is -1.
mat((a,b),(c,d))$ assigns a 2 by 2 matrix.
det(A)$ computes the determinant of A.

cos(x)$ displays cosine x.

---

off echo$ disables the echo from the system.

off nat$ allows us to use the output as an input.

out t$ redirects the output to the terminal. (Default).

---

in "filename"$ reads the contents of the file, filename, into REDUCE.

out "filename.out"$ redirects the output to the file, filename.out

write "comment"$ writes your comment to the output file, filename.out .

shut "filename.out"$ closes the file, filename.out .

---

sub({x=3},y=x)$ substitutes 3 for x in the expression y=x .

mkid(result,3)$ creates a new variable "result3".

solve({f1,f2,f3})$ solves a given system of equations f1=f2=f3=0 .

on fullroots$ attempts to express very explicitly.

length(result)$ reports the number of elements of he list, result .

part(result,j)$ takes the j-th element out of the list, result .

coeffn(f,j)$ takes the j-th coefficient of the polynomialf in one variable.

---

setmod 3$ on modular$ computes in modulo 3.

off modular$ turns off modular computation.

---

on rounded$ starts numerical computation.

precision 50$ adjusts the number of digits for numerical computation to 50.

off rounded$ ends numberical computation. (Default).

REMARK B.2. *The following commands are needed for any computation related to Groebner basis of a polynomial ideal.*

load_package groebenr$ starts groebner package.

torder({x,y,z}, lex)$ assigns the lex order. (Default).

torder({x,y,z}, gradlex)$ assigns the graded lex order.

torder({x,y,z}, revgradlex)$ assigns the reverse graded lex order.

groebner({f1,f2,f3}); compute the reduce groebner basis for the ideal generated by f1, f2, and f3.

---

groebnert({given_f1=f1,given_f2=f2,given_f3=f3}); not only computes the reduced Groebner basis but also keep track of all the intermediate division process so that the output of this command allows us to figure out the quotients of the division by f1, f2, and f3.

groebprotfile; contains all the intermediate division results.

---

greduce(f,{f1,f2,f3}); computes the unique remainder of the division of f by the reduced Groebner basis for the ideal generated by f1, f2, and f3.

preduce(f,{f1,f2,f3}); computes a non-unique remainder of the division of f by f1, f2, and f3.

REMARK B.3. *The following table is sometimes very useful.*

Built In Infix	**Equivalent Expression**
:=	setq
=	equal
>=	geq
>	greaterp
<=	leq
<	lessp
+	plus
-	difference (if unary, minus)
*	times
/	quotient (if unary, recip)
^ or **	expt (raising to a power)
.	cons

REMARK B.4. *For detailed usage informations on these commands, one may visit the following homepage:*

http://trinitas.mju.ac.kr/intro2reduce.html ,

where one will easily find several manuals for the Computer Algebra System, REDUCE.[1]

2. The Linear Algebra Package in MAPLE

To bring a package for Linear Algebra, type in at MAPLE prompt,

with(linalg); (and then press Enter key)

Then you are ready to use MAPLE's linear algebra package, linalg. In MAPLE, vectors and matrices are treated as the same kinds.

2.1. MAPLE commands for Matrices. Once you type in with(linalg); you will see a list of MAPLE commands available from the linalg package, and here are the list:

```
> with(linalg);
Warning, the protected names norm and trace have been redefined and unprotected
[BlockDiagonal, GramSchmidt, JordanBlock, LUdecomp, QRdecomp,
Wronskian, addcol, addrow, adj, adjoint, angle, augment,
backsub, band, basis, bezout, blockmatrix, charmat, charpoly,
cholesky, col, coldim, colspace, colspan, companion, concat,
cond, copyinto, crossprod, curl, definite, delcols, delrows,
det, diag, diverge, dotprod, eigenvals, eigenvalues,
eigenvectors, eigenvects, entermatrix, equal, exponential,
extend, ffgausselim, fibonacci, forwardsub, frobenius,
gausselim, gaussjord, geneqns, genmatrix, grad, hadamard,
hermite, hessian, hilbert, htranspose, ihermite, indexfunc,
innerprod, intbasis, inverse, ismith, issimilar, iszero,
```

[1]http://ch.catholic.or.kr/pundang/archive/4/intro2reduce.html

jacobian, jordan, kernel, laplacian, leastsqrs, linsolve, matadd, matrix, minor, minpoly, mulcol, mulrow, multiply, norm, normalize, nullspace, orthog, permanent, pivot, potential, randmatrix, randvector, rank, ratform, row, rowdim, rowspace, rowspan, rref, scalarmul, singularvals, smith, stackmatrix, submatrix, subvector, sumbasis, swapcol, swaprow, sylvester, toeplitz, trace, transpose, vandermonde, vecpotent, vectdim, vector, wronskian]

Among these commands, we explain several of them very briefly:

(1) matrix commands allows you to input a matrix,
(2) addrow, swaprow, and mulrow commands are for elementary row operations,
(3) inverse command computes the inverse of a matrix,
(4) det command computes the determinant of a matrix,
(5) &* is for matrix multiplication,
(6) evalm is for matrix evaluation,
(7) eigenvals, eigenvalues, eigenvectors, and eigenvects commands are for computation of eigenvectors and eigenvalues,
(8) nullspace command is for the nullspace of a matrix,
(9) colspace command is for the column space of a matrix,
(10) row space command is for the row space of a matrix,
(11) gausselim command is for the gaussian elimination of a matrix, etc.

2.2. Creating matrices.

EXAMPLE B.1. *Here are some examples:*
with(linalg);
A:=matrix([[1,2],[3,4]]);

$$A := \begin{bmatrix} 1 & 2 \\ 3 & 4 \end{bmatrix}$$

inverse(A),det(A);

$$\begin{bmatrix} -2 & 1 \\ 3/2 & -1/2 \end{bmatrix}, -2$$

B:=matrix(2,2,[w,x,y,z]):evalm(a+b);

$$\begin{bmatrix} 1+w & x+2x \\ 3+y & 4+z \end{bmatrix}$$

*evalm(A&*B);*

$$\begin{bmatrix} w+2y & x+2z \\ 3w+4y & 3x+4z \end{bmatrix}.$$

2.3. Accessing matrix components.

EXAMPLE B.2. *Here are some examples:*
with(linalg);
M1:=matrix([1,2,3,4],[a,b,c,d],[e,f,g,h],[w,x,y,z]]);

$$M1 := \begin{bmatrix} 1 & 2 & 3 & 4 \\ a & b & c & d \\ e & f & g & h \\ w & x & y & z \end{bmatrix}$$

M1[2,3];

$$c$$

submatrix(M1,1..2,3..4);

$$\begin{bmatrix} 3 & 4 \\ c & d \end{bmatrix}$$

subvector(M1,2,1..4);

$$[a, b, c, d]$$

row(M1,4);

$$[w, x, y, z]$$

col(M1,2);

$$[2, b, f, x]$$

2.4. Elementary arithmetic operations with matrices.

EXAMPLE B.3. *Here are some examples:*

with(linalg);

M1:=matrix([[1,2],[3,4]]): transpose(M1);

$$\begin{bmatrix} 1 & 2 \\ 3 & 4 \end{bmatrix}, \begin{bmatrix} 1 & 3 \\ 2 & 4 \end{bmatrix}$$

evalm(1/M1);

$$\begin{bmatrix} -2 & 1 \\ 3/2 & -1/2 \end{bmatrix}$$

map(sqrt, M1);

$$\begin{bmatrix} 1 & \sqrt{2} \\ \sqrt{3} & 2 \end{bmatrix}$$

*evalm(M1^3), evalm(M1&*M1&*M1);*

$$\begin{bmatrix} 37 & 54 \\ 81 & 118 \end{bmatrix}, \begin{bmatrix} 37 & 54 \\ 81 & 118 \end{bmatrix}$$

2.5. Standard Examples in Linear Algebra.

EXAMPLE B.4. *Here are some examples:*

 # The following line invokes each of the inputs read in from a file to be displayed.

interface(echo=2);
with(plots):
with(plottools):
with(linalg):
*a:=matrix([[1,0,0,0,1,0,0,0],[5,2,0,0,0,1,0,0], *
[8,6,3,0,0,0,1,0],[0,9,7,4,0,0,0,1]]);
a1:=addrow(a,1,2,-5);
a2:=addrow(a1,1,3,-8);
a3:=addrow(a2,2,3,-3);
a4:=addrow(a3,2,4,-9/2);
a5:=mulrow(a4,2,1/2);
a6:=mulrow(a5,3,1/3);
a7:=addrow(a6,3,4,-7);

```
a8:=mulrow(a7,4,1/4);
aa:=submatrix(a,1..4,1..4);
b:=submatrix(a8,1..4,5..8);
check:=evalm(aa &* b);
b1:=vector([2,3,4,5]);
result:=linsolve(aa,b1);
check2:=evalm(aa &* result);
# The following line selects different plotdevice other than the default one.
# Using ghostview command at unix % prompt, you can display the graph of sin(x).
# interface( plotdevice=x11, plotoutput=myfile );
interface( plotdevice=ps, plotoutput=myfile );
p1:=plot3d(sin(x)*cos(y),x=-2*Pi..2*Pi,y=-2*Pi..2*Pi):
p2:=plot3d(x*y,x=-2*Pi..2*Pi,y=-2*Pi..2*Pi):
display({p1,p2});
# For instance, ghostview myfile (enter) at unix % prompt will display the graph of sin(x).
interface(plotdevice=default);
```

EXAMPLE B.5. *Here are some other examples:*

```
# The next line allows maple to disply what you typed in:
interface(echo=2);
with(linalg);
m1:=matrix([[a,b,c],[d,e,f],[g,h,i]]);
v1:=vector([u,v,w]);
v2:=matrix([[u],[v],[w]]);
evalm(m1 &*v1);
innerprod(v1,v1);
innerprod(v1,m1);
A:=matrix([[1,2,3],[4,8,12],[1,-1,0]]);
b:=vector([14,12,46]);
linsolve(A,b);
```

```
kernel(A);
nullspace(A);
eigenvals(A);
eigenvects(A);
poly:=charpoly(A,'x');
solve(poly,x);
v1:=vector([-1,2,3,0]);
v2:=vector([0,1,2,1]);
# v3:=vector([2,-1,-1,1]);
v3:=evalm(2*v2);
basis( {v1, v2, v3} );
GramSchmidt([v1,v2,v3],normalized);
```

2.6. How to Read MAPLE Codes. MAPLE is a procedural programming language, like C. With the exception of the commands belonging to the kernel and the compiled library commands, we can view the code of all MAPLE commands with 'print'.

EXAMPLE B.6. *Here are some examples:*

```
#To review the MAPLE command 'det' in Linear Algebra package,
with(linalg):
interface(verboseproc=2);
A:=mat([[1,2],[2,1]]);
det(A);
print(det);
# To review the MAPLE command 'charpoly' in Linear Algebra package,
with(linalg):
interface(verboseproc=2);
A:=mat([[1,2],[2,1]]);
charpoly(A,x);
print(charpoly);
```

APPENDIX C

Suggestions for Further Reading and References

M. Artin, *Algebra*, Prentice Hall, 1991.

G. Birkhoff and S. MacLane, *A Survey of Modern Algebra*, 4th Edition, Macmillan, New Nork, 1997.

D. Cox, J. Little, and D. O'Shea, *Ideals, Varieties, and Algorithms*, 2nd Edition, UTM, Springer-Verlag, 1997.

J. Fraleigh, *A First Course in Abstract Algebra*, 7th Edition, Addison-Wesley, 2000.

A.C. Hearn et al, *REDUCE User's Contributed Package Manual*, Version 3.8, Konrad-Zuse-Zentrum, Berlin, 2004.

T. Hungerford, *Algebra*, Graduate Text in Mathematics, Springer-Verlag, 1974.

G. Kempf, *Algebraic Varieties*, London Math. Soc. LNS 172, Cambridge Univ. Press, 1993.

G. Kempf, *Algebraic Structures,* Friedrich Vieweg & Sohn, 1995.

O.T. Kim and S.A. Park, *Modern Algebra* (Korean), 3rd Edition, Cheongmungak, 1993.

M. Kofler, *Maple: An Introduction and Reference*, Addison-Wesley, 1997.

S. MacLane and G. Birkhoff, *Algebra*, 2nd Edition, Macmillan, New York, 1979.

S.T. Soh, *Algebraic Analysis I*, Kyowoo, Seoul, 2006.

B.L. van der Waerden, *Modern Algebra*, Ungar, New York, 1970.

O. Zariski and P. Samuel, *Commutative Algebra,* Vol.1 and 2, Springer-Verlag, 1958.

http://trinitas.mju.ac.kr/intro2maple.html

http://trinitas.mju.ac.kr/intro2reduce.html

http://trinitas.mju.ac.kr/intro2InetCompu.html

http://ch.catholic.or.kr/pundang/archive/4/soh/2002_InetCompu

http://www-groups.dcs.st-and.ac.uk/~history/

http://www.reduce-algebra.com/docs/groebner.pdf

http://www.reduce-algebra.com/docs/reduce.pdf

Index